Thoughts on Hospital Design and Construction in China

Lun Ge · Zhe Wang

Thoughts on Hospital Design and Construction in China

中国建筑工业出版社
CHINA ARCHITECTURE & BUILDING PRESS

Springer

Lun Ge
Beijing University of Civil Engineering
and Architecture
Beijing, China

Zhe Wang
Department of Architecture
Henan University
Kaifeng, China

ISBN 978-981-97-0817-8 ISBN 978-981-97-0818-5 (eBook)
https://doi.org/10.1007/978-981-97-0818-5

Jointly published with China Architecture & Building Press
The print edition is not for sale in China (Mainland). Customers from China (Mainland) please order the print book from: China Architecture & Building Press.

This Springer imprint is published by the registered company Springer Nature Singapore Pte Ltd.
The registered company address is: 152 Beach Road, #21-01/04 Gateway East, Singapore 189721, Singapore

Contributors

Lun Ge and Zhe Wang contributed equally to this book. The interviewees were Haixiao Chen, Jie Qiu, Yunxiang Liu, Roberta Lipson, Yingkang Shi, Chen Wang, Lirong Zhuge, Jianzhong Zhang, Mei Chen, Jianhua Dong, Rong Ma, Xin Ni, Chongde Shen, Tielin Wang, Wei Zhang, Qixia Zhao, Yi Zheng, Guoliang Chen, Jian Gu, Xiqiu Huang, Jianmin Meng, Xi Zhu, Vincent Zhang.

These interviewees contributed equally.

This book was written and reproduced from the individual interviews with these interviewees by Lun Ge and Zhe Wang with permission from the interviewees.

Foreword

It is a pleasure to recommend the English version of this interview collection related to design and construction of hospitals in China. It offers a fascinating and useful insight to the perspectives of administrators, facility executives, physicians, and architects as they consider hospital design in China. The topic should be of interest both to the field of hospital design as practiced in China and to international practitioners who hope to collaborate on projects in China. Much of hospital design around the world is universal, as in the case of infection prevention, but the means of achieving design goals and the need to respond to national and regional differences require an understanding of the hospital context. The Chinese government has announced its commitment to increase the number of hospitals serving its population. As Chinese and international organizations approach a need to design these important facilities, they would be well served to review the content of these interviews in order to better understand the multiple perspectives of these deeply involved professionals. I congratulate the authors, interviewers, and participants who have contributed to this important and welcome publication.

TX, USA D. Kirk Hamilton

Preface by Lun Ge

I graduated from college in 1987 from Harbin Institute of Technology; I chose design for a general hospital building as the project required for graduation. From then on, my career has tied up with design for hospitals. From 1996 to 1999, I was honored to be chosen to be a trainee and got Master degree in hospital architectural design in the Leuven University in Belgium. During the intensive study, I was deeply attracted by the unique charm and unfathomable mystery for hospital architectural design so much so that I decided to embark on the worthy but challenging path of exploration for hospital architecture in my career later. It was a great honor to study under the supervision of Prof. Yan Delrue, an established architecture in the field of hospital architecture in Europe. He has a profound understanding of the theory of hospital architecture and a unique methodology. I benefited a great deal in my career from him.

From 2001 to now, I have focused on the education, research, and practice of hospital construction for more than 20 years. My team has obtained the grants for the research project of the National Natural Science Fund for architecture and the key research project of Beijing Natural Science Fund. It is our team which, for the first time in China, proposed "Planning theory and method of medical facilities" and "The theory and method for coupling of space and function as double-stranded DNA". Besides, our team developed the "SHAPE" as the protocol for POE. Figure 1 presents the concept of POE.

We summarized the three field research contributions as the term "three carriages," which is the most important and indispensable component of hospital construction. I have kept calling for the recognition of the "three carriages" from the government and hospital clients as the closed-loop working approach. This effort serves as the original intention of this book. Figure 2 presents the concept of "three carriages."

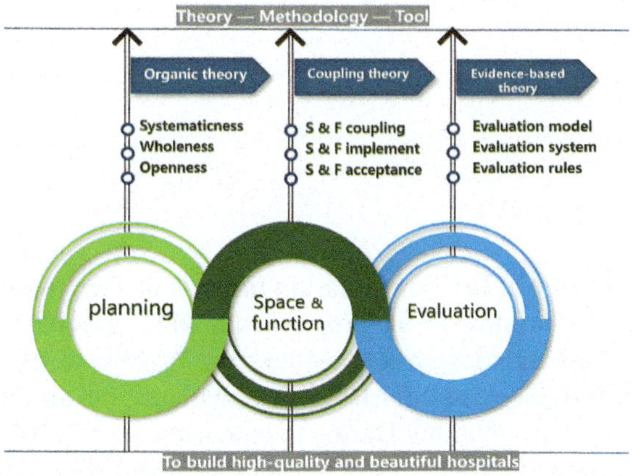

From the mid-90s to the beginning of the twenty-first century, hospital construction in China reached an unprecedented peak. Almost all the provincial and metropolitan-level hospitals have undergone new constructions and/or expansions. Due to the rapid and excessive amount of construction, many problems were reflected in the use. From 21C to the 2010s, it was time to reflect on what was indeed to clarify the causes of these problems. From 2020s, the National Health Commission put forward the initiative of building high-quality and beautiful hospitals, which we must achieve the "Three Improvements": the improvement in efficiency, in quality, and in remuneration. We must realize the "Three Characterizations": humanization, functionalization, and intelligenization. Figure 3 presents the history of hospital construction in China.

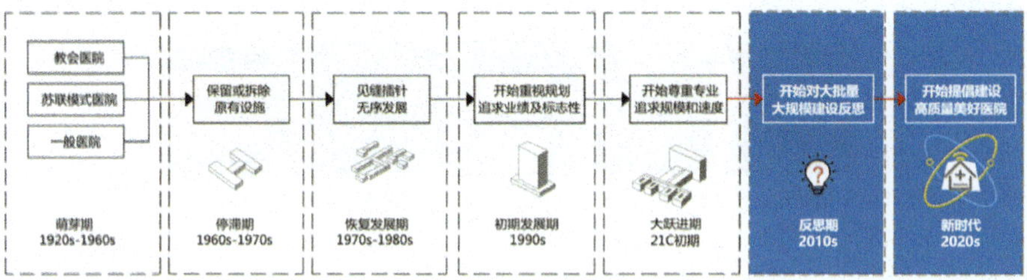

At the same time, we have also deeply realized that the Internet, big data, and other technology have brought such profound changes as in the management and construction for hospitals. The future has come—artificial intelligence and Internet will play an immeasurable role in the development and construction of Chinese hospitals in the future.

From January 2014 to January 2015, my team and I have traveled to different places in China. We have interviewed, face-to-face, and in-depth 45 specialists in medical construction.

Every specialist, in his or her respective field, has accumulated rich experiences and offers his or her unique insights. The interviews reflect true and most thoughtful rational voice in the field of contemporary hospital construction. In May 2015, the Book—"Thoughts on Chinese Hospital Architecture—Gelun's Interviews" was officially published in Chinese. In this book, we rewrote the records of selected 23 interviews from a total of 45, which would be of interest to the readers from foreign countries. We hope that the readers will feel the contents are meaningful. Hopefully, the book can serve as a broader communication platform and bring in social benefits.

I am taking this opportunity to salute all the interviewees. Time flies. The Year of 2023 is approaching the end of the year. I wish for the coming of a new and better era in the field of the construction of medical facilities in China.

Beijing, China
October 2023

Lun Ge

Preface by Zhe Wang

Hospital Design and Construction in China and the Interview Records

Many designers engage in international practice in China and are curious about how to better navigate their design projects in this country? Regarding hospital design and construction, the role of local cultures is often more important than expected and thus demands careful consideration. In this field, many issues have been discussed in conferences or meetings but limited information has been published.

Thoughts on Hospital Design and Construction in China provides an answer to the above question. It illustrates the insights collected via interviews with eight hospital presidents, nine vice presidents focusing on construction management in hospitals, and six hospital architects. These interviews were originally conducted in 2015 by Lun Ge. We reiterated the thoughts with the interviewees in 2023 and rewritten the records with updates to reflect current perspectives.

One of the distinguishing features of this book is its use of "real" stories to interpret the landscape of hospital design and construction in China. Internationally, we highly recommend evidence-based design research for design decision making. But how pervalent is design research in China? It is clear that the incorporation of research into design is gaining recognition in China. The ideology of a hospital, processes of building design and construction, and healthcare policies in China are interpreted in this book. Designers and practitioners can use the information to investigate indigenous local user needs in the specific context of their projects.

This book offers a comprehensive overview of the current state of hospital design and construction in China. Experts' thoughts regarding design decision making and the supervision systems for hospital development are discussed. In order to win, designers must consider the cultural implications of studies for designing. Those with a better understanding of user needs and client requirements are more likely to keep and increase their market share.

Kaifeng, China
October 2023

Zhe Wang

Contents

About the Authors

Professor Lun Ge has engaged the work of hospital architecture in the "Industry-Education-Research" fields for more than 20 years, put forward the original theory of "the three troikas" (PPE: whole Planning-whole Program-whole Evaluation). Based on the theory, the team under her leading independently developed GMARS methods for preliminary planning, the guideline for medical program coupling design and SHAPE + evaluation tool, which has "consulted" for 100 + hospitals, solved 1000 + problems, saving 5%–10% of the total investment in hospital construction, and making outstanding contributions to the high-quality development of the whole life cycle of hospital building in different stages; which fills the gap of theory and method in the field of hospital construction. Professor Lun Ge became the first person of the "Before planning and after evaluation" for the hospital construction in China. The team under her leading has undertaken the consulting services of hundreds of hospitals, won the approval of the hospital clients, and the design project won the national Gold medal.

Published monographs:

Thoughts on Hospital Construction in China: Gelun Interview Records
Centennial Old Building of PUMC 1921–2021 I and II volume

1. Beijing University of Civil Engineering and Architecture Professor, Chief Expert of the Healthcare Architecture Research Center;
2. Shenzhen University Benyuan Design and Research Center, Honorary Director of the Healthcare Department;
3. National Health Commission of the P R C, Beijing Hospital Authority, Hospital Construction Consultant;
4. China Health Culture Association, Vice President for Healthcare Division;
5. Gerontological Society of China, Vice President for Healthcare Division;
6. International Academy for Design and Health, Vice President of China Division.

Zhe Wang, Ph.D., RA, EDAC, LEED AP is the holder of Yellow River Professorship at Henan University and titled "Distinguished Professor" by Henan Province, China. She received her Ph.D. degree and Master of Architecture degree from Texas A&M University and was Principal Investigator of Research at Cannon Design, USA. Responding to the needs of patients and older adults, Wang's research has focused on environments for health. She has published 3 books and 39 first-author articles in peer-reviewed journals, including the *Journal of Health and Place, Gerontologist, Transport and Health, Healthcare Environments Research and Design Journal*, and *AIA Reports.* Her research "From Pre-design to Post Occupancy Evaluation—Cancer Treatment Environments" received the highly commended International Academy Awards in Research, announced by the World Health Design following a two-stage judging process involving more than 25 world-renowned researchers. As the committee chair of 2023 Great Places Awards, Wang has served the Environmental Design Research Association (EDRA) since 2010 and co-developed the EDRA CORE program.

Chapter 1
Essentials at a Glance

Interviews with Hospital Presidents

Create a Hospital with Warm Atmosphere for Patients—An Exclusive Interview with Haixiao Chen

Haixiao Chen: President of Zhejiang Taizhou Enze Medical Group, Professor, Chief Physician, and Doctoral Supervisor; Deputy to the 10th, 11th, 12th National People's Congress; Awardee of the National Labor Medal, the Award of Best National Hospital Director, and the Zhejiang Science and Technology Progress Awards.

Key sentences:

1. *If I were the president, I wouldn't go for an expensive building. I may not be able to build the best building but a building that suits our hospital the best.*
2. *In my opinion, one issue of the hospital design and construction in China is that everyone values hospital planning but no one knows how to plan...Emphasizing the importance of user experience is the core of hospital design.*
3. *I remind myself from time to time: Be aware of what you can do and what you cannot. I can cure patients but I can't build buildings. However, I need to build buildings for our hospital, since me as the president can't trust others to do it. Buildings must be constructed for a hospital and they set up the platform for hospital development. A hospital president should firstly identify people who can build buildings and secondly strengthen self-learning to understand at least the basic and principal knowledge of building design and construction. In this regard, we lack an architectural knowledge training system for hospital presidents or managers who are in charge of hospital development in China.*

Design a Hospital with Soul—An Exclusive Interview with Jie Qiu

Jie Qiu: President of Gansu Provincial Maternity and Child-care Hospital, Chief Physician, Professor, PhD Supervisor; 1st-level Leading Talents of Gansu Province, Awardee of the National Labor Medal and the Excellent Hospital President in China.

L. Ge and Z. Wang, *Thoughts on Hospital Design and Construction in China*,
https://doi.org/10.1007/978-981-97-0818-5_1

Key sentences:

1. *Today's hospital development shouldn't be focused on the piling up of expensive building materials, but help people to get the feeling of returning to nature and having the inner harmony with nature, leading to stimulated self-healing.*
2. *The first task is to improve the quality of populations. The second is to improve the quality of life. Therefore, the "soul" of our hospital is to contribute to these responsibilities and duties entrusted by our people.*
3. *I hope that forward-looking building standards and norms for maternal and child care can be well developed in line with China's reality. The standards should be oriented and also a support point for promoting the quality of populations. This is by no means building a maternal and child-care hospital can solve all the problems, but the responsibility should be carried out and promoted by the hospital.*

Making Functional Master Plans—An Exclusive Interview with YunXiang Liu

YunXiang Liu: Former Party Secretary and Deputy Director of Health Bureau Yantai, Shandong. Former Party Secretary and President of Yuhuangding Hospital, Yantai. Holder of the honorary titles "National Model Worker", "Shandong Province Model Worker", "National Outstanding President." Awardee of the "National Outstanding Contribution Award in Hospital Management" and the "National Hospital Construction Lifetime Achievement Award."

Key sentences:

1. *Regarding hospital management, I think the key is having a theory of development and a good understanding of the functional flows and their relationship.*
2. *Hospital managers shouldn't rely on architects and planners to design a dream hospital since these designers aren't medical professionals. Designers typically design based on a design concept. As medical professionals, hospital managers should be involved in the design of medical service processes, department functions, and design details.*
3. *The key for hospital planning, design, and management is to promote the process of medical services. Integrated planning and humanization design from the standpoint of patients are preferred.*

Persistence on Original Intents—An Exclusive Interview with Roberta Lipson

Roberta Lipson: Founder and Vice Chair, United Family Healthcare, New Frontier Health

Key sentences:

1. *Our mission has always been first and foremost to deliver the best medical care to our patients, and we believe that if we achieve that profits follow.*
2. *The core of high-end healthcare service is respect for patients in every way. It is a type of spirit and culture that come from the inside and must be internalized by every employee.*

3. *China's healthcare industry is still in transition. We hope UFH can present a new mode and concepts of healthcare. Together with the refinement of policies and regulations, we are committed to constant improvement of healthcare safety and service standards.*

Optimize the Distribution and Manipulation of Medical Resources—An Exclusive Interview with Yingkang Shi

Yingkang Shi: Professor; Committee member of the PhD programs on cardiovascular and thoracic surgery, hospital administration, and medical informatics; former President of the West China Hospital affiliated with Sichuan University.

Key sentences:

1. *Our top task is to make effective use of the limited high-quality medical resources available to us. Another issue is the limited "accessibility" of the medical resources. Accessibility should be improved to promote the efficiency of medical services. It requires efficient utilization of devices and spaces."*
2. *As a matter of fact, I think a preliminary planning of the overall project and the considerations on user needs are of the most importance. Moreover, during a hospital planning, we should take the situation of disasters into account since hospitals shall by no means stop saving patients in case of disasters.*
3. *As time goes by, there will be less resources worldwide and thus the global situations will be more severe: higher GDP, population aging, incurable chronic and geriatric diseases, and increased medical expenses. These are issues worldwide.*

Open Mind, Update Concepts and Ideas, and Make Good Standards—An Exclusive Interview with Chen Wang

Chen Wang: President of Tiantan Hospital affiliated with Capital Medical University; Vice President of Beijing Tongren Hospital affiliated with Capital Medical University; Vice President of the Beijing Hospital Association.

Key sentences:

1. *From different perspectives, we both give priority to the importance of service processing in design decision making and agree that the convenience for patients and physicians is the core of design.*
2. *Of every factor, including technology and construction, the norms and codes should be kept updating. The standards of hospital construction must be adapted to the state-of-art of medical services and technology.*
3. *Indoor air is more important than the outdoor air...It is very necessary to improve natural ventilation in hospital design. It is a key aspect that must be considered for medical buildings in the future.*

The Soul of Hospital Administration—An Exclusive Interview of Lirong Zhuge

Lirong Zhuge: Senior economist; former Executive Associate Director of Shanghai Proton and Heavy Ion Center Preparatory Office; former Associate Director of Shanghai ShenKang Hospital Development Center; former Chairman of the Hospital

Architecture System Research Branch of Chinese Hospital Association; Director of Institute of Fudan Hospital support service & management.

Key sentences:

1. *If a construction plan were issued without specific standards, it would have negative influences on project management, leading to excessive scales, standards, and investments (The Three Excesses). These would be an issue of government management.*
2. *We advocate natural lighting and ventilation and it should be considered on a case-by-case basis.*
3. *Many international companies provide beautiful design renderings to attract eye balls, but aren't address the needs of building functions well.*

Building an Integrated System for High-Quality Hospitals—An Exclusive Interview with Jianzhong Zhang

Jianzhong Zhang: Senior Economist, National Certified Supervisory Engineer, Royal Chartered Surveyor, US Certified Engineering Cost Engineer, Director of Shanghai Hospital Construction Management Center, CEO of Shanghai Shenkang Hospital Development Center (SHDC).

Key sentences:

1. *We shall "indigenize" experience from foreign countries and conduct research and exploration in the context of China.*
2. *Promoting the management mechanisms enhances the quality of operation. Operations conducted on a "large platform" facilitate monitoring and respond to future inquiries.*
3. *It is necessary to integrate and assemble the factors of pre-design planning and post occupancy evaluation …We need to formulate a good mechanism of management. It's necessary to gain a thorough understanding of a hospital's organizational background and its construction systems.*

Interviews with the Vice Presidents Focusing on Construction Management in Hospitals

Hospital Construction is a Range of Controls from Fundamental Construction to Logistics Management—An Exclusive Interview with Mei Chen

Mei Chen: Vice president of Shanghai Sixth People's Hospital, Senior engineer, Deputy Secretary General of China Hospital Association's Building System Research Branch, Vice president of the Logistics Management Research Institute of Fudan Hospital.

Key sentences:

1. *It requires a professional background and scientific methods to address user needs and meet the requirements for project continuity, security, energy saving, consumption reduction, and less or no equipment breakdowns.*

2. *Preliminary planning and POE are crucial to hospital construction. In the overall management, we should hold a holistic view on various factors associated with the preliminary planning, project verification, and construction documents.*
3. *When a hospital is occupied, the quality of equipment, such as valves of pumps, catches attention. Once there is any equipment breakdown, its influence can be fatal to the hospital. Therefore, during the construction, we should pay attention to the equipment, make sure it meets the standards for energy saving and safety. Construction and practical operation should be considered simultaneously.*

Hospital Design Starts from a Comprehensive "Diagnosis"—An Exclusive Interview with Jianhua Dong

Jianhua Dong: Former Director of the Construction Department of Peking University's First Hospital, Vice President of Beijing Hospital Construction Association.

Key sentences:

1. *Project positioning is the foundation of everything.*
2. *In fact, evidence-based research in early design is necessary in order to identify appropriate medical technologies and procedures for the design. Otherwise, it's essentially impossible to identify them.*
3. *It's necessary for a hospital to make a thorough and indigenized analysis of itself, leading to a clear understanding of its future development.*

To Be a Good Translator and Interdisciplinary Facilitator—An EXCLUSIVE interview with Rong Ma

Rong Ma: Vice president of the Drum-tower Hospital in Nanjing.

Key sentences:

1. *Successful hospital design and construction relies on neither the designer nor the hospital itself. It requires interdisciplinary cooperation.*
2. *Translators and interdisciplinary facilitators are important in hospital design and construction.*
3. *Upon designing a hospital, one must consider how the hospital systems could be run efficiently. So, the key to developing an efficient hospital lies in design.*

Tackling Architectural Challenges with "Anatomy"—An Exclusive Interview with Xin Ni

Xin Ni: Chief Physician, Professor, Doctoral Supervisor; Director of National Center for Children's Health; President of Beijing Children's Hospital affiliated with Capital Medical University; President of Beijing Pediatric Research Institute; Dean of the School of Pediatrics at Capital Medical University.

Key sentences:

1. *Being a medical doctor, I am specialized in human anatomy. Similarly, you architects are also "medical doctors", who work on the anatomy of buildings. My work on human anatomy may be more complicated than yours. Now that I've done*

perfectly in such a delicate domain of human anatomy, the 'anatomy of medical architecture' wouldn't be intractable to me. So I hope you can trust me and let us work together to get the project done successfully.

2. *Never undervalue the importance of hospital logistics and the logistics personnels. These support a hospital' infrastructure. A hospital cannot function well without a safe built environment. We should learn from all aspects to gain valuable experiences in practice.*
3. *No matter about its type, a hospital should aim at completing diagnosis and treatment services through the shortest pathway and in the shortest time.*

Building Hospitals: Blessings Bestowed by Our Practice—An Exclusive Interview with Chongde Shen

Chongde Shen: Vice President of Wuxi People's Hospital, Supervisor of Masters' and Doctor's programs. Chief Member of the Standards Professional Committee of China Health Information Association; Vice Chairman of the Information Statistics Professional Committee of China Hospital Association.

Key sentences:

1. *In many projects, on the first day of occupancy, building renovation starts. These problems result from deficient systematisms and professionality.*
2. *Traditional Chinese Medicine hospitals promote Chinese cultures as well as Chinese medical cultures. TCM is the quintessence of Chinese cultures and offers the most concentrated expression of Chinese philosophy.*
3. *Although we work in the field of hospital logistics and don't directly face the patients, we actually indirectly provide services to them and should be dedicated to a great love and with valuable recognition of the job that is to support patients and their families, and also to let the employees have a correct understanding of themselves. We help them to understand the uniqueness of their work. It promotes universal love.*

Solving Professional Problems Professionally—An Exclusive Interview with Tielin Wang

Tielin Wang: President of Hainan Cancer Hospital, former Vice President of TEDA International Cardiovascular Hospital in Tianjin, former President of Tianjin TEDA Hospital, Member of the hospital construction expert panel of the Health Ministry of China, President and Member of the Building Committee of Chinese Hospital Association, Drafter of the *Design Specifications for General Hospital Buildings, Construction Standards for General Hospitals, Construction Standards for the Clean Operating Departments of General Hospitals*, and *Technical Specifications for the Clean Operating Departments of General Hospitals*, Editor of the *Chinese Hospital Management Encyclopedia, Guidelines on Hospital Surgical Department Construction.*

Key sentences:

1. *A hospital should be an organic system. The deeper we investigate this system, the more we recognize the importance of utilizing integrated spaces. Every square meter should be used appropriately.*
2. *We lack medical planners and need professional designs, construction, and consultancy. A project review would be meaningful if it improved the project and guided its future development.*
3. *It wouldn't be a matter if all my words got deleted from this book but saving a single sentence saying "Technical processing of functional flows is the foundation of hospital design".*

New Administration Patterns for Future Hospitals—An Exclusive Interview with Wei Zhang

Wei Zhang: Executive Vice President of West China Hospital of Sichuan University.

Key sentences:

1. *According to Sun Tzu's Art of War, "Food and fodder should go ahead of troops and horses". Regarding hospital administration, we value not only the support of logistics to healthcare services, staff education, research, and administration, but also the positive influences of good logistics on hospital development, especially the facilitation and promotion of healthcare disciplines.*
2. *Future hospital logistics is a broader concept of 'general logistics'. It requires the integration of all departments, refinement of duties, and sorting of procedures. The 'position' or role of an individual staff member would be more important than that of the department to which he or she were assigned.*
3. *Informatization is an important part of logistics management in which standardization, formulization, and informatization are interrelated with each other.*

Steadfast Work and Focus on Data—An Exclusive Interview with Qixia Zhao

Qixia Zhao: Division Head of the Health Facility Management Department at Peking University Third Hospital.

Key sentences:

1. *In POE, hospital architects need to investigate how to design clear space and functional programs for medical services, follow the current standards, and thus promote hospital development.*
2. *Quantitative research regarding hospital construction needs to be supported by specific data, which in turn can provide evidence for POE. Using the quantified findings from data analysis can help to improve the quality of design.*
3. *Healthcare architects should pay more attention on the development of outpatient departments and medical technologies, and thus can objectively determine the appropriate proportions of floor areas for the major departments in a hospital.*

Respect the Basic Rules of Construction, Make Excellent Planning and Management—An Exclusive Interview with Yi Zheng

Yi Zheng: Former Assistant-President of Beijing Ditan Hospital affiliated with Capital Medical University.

Key sentences:

1. *The first lesson we have learned is that we must respect design and the basic rules of construction and leave a plenty of time for preliminary planning and argumentation.*
2. *It is very necessary to spend more time in preliminary planning, in order to make the planning detailed, the argumentation sufficient, and the proposed functions complete.*
3. *Renovating a hospital was even harder than building a new one. In order to completely solve the problems, we need a complete reform of our hospital. We should have foresight in project planning, project scale, investment, and many other aspects in order to build a truly sustainable hospital.*

Interviews with Hospital Architects

Exploring in a Complicated Situation—An Exclusive Interview with Guoliang Chen

Guoliang Chen: Chief architect of Shanghai Architectural Design & Research Institute Co., Ltd.

Representative work: the Xiamen Branch of Zhongshan Hospital affiliated with Fudan University, Fujian Province.

Key sentences:

1. *I think the functionality of a hospital street should be preferred than its symbolic value. It serves as a hub connecting the individual functional areas for medical services. It's part of the transportation system, which may have diversified forms.*
2. *It would be quite problematic if hospital design and construction were not carried out either from the inside to the outside or by integrating the interior and exterior design.*
3. *If we invited oversea designers for hospital projects, they should remain in the team during the schematic design phase, complete the primary and secondary level of design on functional flows. What we need is their experience designing the hospital as a whole project rather than a mere shell.*

Obligations of Hospital Designers—An Exclusive Interview with Jian Gu

Jian Gu: Chief architect of China IPPR International Engineering Co., Ltd., Deputy director of the institute for medical architectural design, first class registered architect, senior engineer.

Representative work: Foshan First People's Hospital, Guangdong Province; Foshan Women's and Children's Hospital, Guangdong Province.

Key sentences:

1. *Therefore, if we need to sort out the technical processes for hospital design, we should figure out the flow relationships in accordance with the logic of medical services.*
2. *The brief of a project is critical as it is the basis guiding the design.*
3. *Interior design is not simply about decoration, but involves genuine 'functional designs' in the interior. It should be an extensive part of architectural design. Moreover, interior design influences service modes and involves certain concerns on psychology.*

Medical Architects and Challenges in the Era of Transition—An Exclusive Interview with Xiqiu Huang

Xiqiu Huang: Famous medical architect in China, Winner of the 6th "Liang Sicheng Architecture Award".

Representative work: Foshan First People's Hospital, Guangdong province; Beijing Xiaotangshan Hospital, Beijing.

Key sentences:

1. *Modern hospital design and construction in China has been developing more and more rapidly after the Reform and Open policy issued in China. However, our architects are not well prepared in terms of understanding the policies, design concepts and methods of designing hospitals.*
2. *Regarding the procedures, what we are used to making decisions on the scale and budget before conducting project feasibility study. This sequence causes many problems.*
3. *Due to insufficient development of community-level medical services, it is hard to create a completed hierarchical healthcare system and thus leads to the phenomenon of overcrowded at large hospitals in big cities.*
4. *At the turning point of national development, there are insufficient connections among disciplines, insufficient preliminary planning, and ambiguous positioning of a hospital project. Rational design and development of hospitals is necessary.*

Comprehensive Hospital Design for Humanistic Care—An Exclusive Interview of Jianmin Meng

Jianmin Meng: Member of the Chinese Academy of Engineering; Chief Architect of Shenzhen General Institute of Architectural Design and Research Co., Ltd.; Distinguished Professor of Shenzhen University; Recipient of the 7th Liang Sicheng Architecture Award.

Representative work: Shenzhen Hospital affiliated with the University of Hongkong, Guangdong Province; Zhangjiagang First People's Hospital, Jiangsu Province.

Key sentences:

1. *Without a good planning, the design would be rigid or merely repeating the precedents without any improvement. One of our clients frankly criticized our*

design institute for repetitive design mistakes. He said that our designers kept carrying the same mistakes from one project to the other. This really upset me. Since then, I often refer it to warn our designers. Actually, this is a common problem in many design institutes.

2. *First of all, the most fundamental requirement is to meet the needs for hospital functions, including medical technologies and service processes. Secondly, we have been advocating the concept of all-round care, which embodies the ideology of humanity. In other words, we care all the user groups in a hospital instead of merely the patients. The third point involves the idea of green and ecological development (ventilation, natural lighting, etc.). The fourth one is about technologies and costs. We suggest budgeted design and economic efficiency instead of blindly using expensive products or materials. Fifthly, we should incorporate the ideas of urban and architectural aesthetics in design.*

As Bold as a "Dinosaur" in this Commercial Age—An Exclusive Interview with Xi Zhu

Xi Zhu: Vice president in charge of the design of Bouygues Construction Group (France); BMP, Shenzhen Baipeng Architectural Design Co., Ltd.; Senior engineer who receives special allowance from the government; First-class registered architect.

Representative works: the First Hospital affiliated with Zhengzhou University, Henan Province; the First Hospital affiliated with Guangdong Medical University, National Center for Respiratory Medicine.

Key sentences:

1. *In a hospital project, initial design review is a key step in design assessment, which includes comprehensive assessment of architecture design and technological design. Design institutes revise and improve their designs according to reviewers' comments.*
2. *Construction drawing design, including architecture design, interior design, and specialty projects, need to be done simultaneously since design factors may overlap and interrelated with each other. An executive architect is responsible to administrate and coordinate design as a whole to prevent design factors from separation.*
3. *The "7 Departments" principle in the 2008 Building Codes is vague, whereas projects vary dramatically and thus classifying is difficult. The idea of classifying by system, provides a detailed and suitable method. In a hospital, the proportion of an individual system depends on the hospital. We could gain experience from other fields. For example, we learn the planning of construction from real estate and project classification from hotels.*

Culture and Humanity in Designing the Drum Tower Hospital—An Exclusive Interview with Vincent Zhang

Vincent Zhang: Swiss-Chinese, Registered architect of Swiss Federal Institute of Architects and Engineers (SIA), Founder of Lemanarc SA (an architecture and urban planning design office) in Switzerland, Chief Architect of the Drum Tower Hospital in Nanjing. One of the first Chinese in Switzerland's history to become a registered architect of the Swiss Federal Institute of Architects and Engineers (SIA), pioneer of the ideology of Gardenized Architecture.

Representative work: Nanjing Drum Tower Hospital, Jiangsu Province.

Key sentences:

1. *As a Chinese architect, one should design in the context of local Chinese cultures and then consider culture integration.*
2. *In the entire system of Drum Tower Hospital, functions always come before forms. It reflects the ideology of rational design. The modular composition of "Medical Matrix" in design responds to service functions.*
3. *It's necessary to think about not only how to construct the entire building at a reasonable cost but also how to adapt this with the current construction technologies in China.*

Chapter 2
Interviews with Hospital Presidents

2.1 Create a Hospital with Warm Atmosphere for Patients—An Exclusive Interview with Haixiao Chen

Time: January 2014—reiterated in May 2023
Location: Taizhou Enze Medical Center

Haixiao Chen: President of Zhejiang Taizhou Enze Medical Group, Professor, Chief Physician, and Doctoral Supervisor; Deputy to the 10th, 11th, 12th National People's Congress; Awardee of the National Labor Medal, the Award of Best National Hospital Director, and the Zhejiang Science and Technology Progress Awards (Fig. 2.1).

Key sentences:

1. *If I were the president, I wouldn't go for an expensive building. I may not be able to build the best building but a building that suits our hospital the best.*
2. *In my opinion, one issue of the hospital design and construction in China is that everyone values hospital planning but no one knows how to plan...Emphasizing the importance of user experience is the core of hospital design.*
3. *I remind myself from time to time: Be aware of what you can do and what you cannot. I can cure patients but I can't build buildings. However, I need to build buildings for our hospital, since me as the president can't trust others to do it. Buildings must be constructed for a hospital and they set up the platform for hospital development. A hospital president should firstly identify people who can build buildings and secondly strengthen self-learning to understand at least the basic and principal knowledge of building design and construction. In this regard, we lack an architectural knowledge training system for hospital presidents or managers who are in charge of hospital development in China.*

Lun Ge: As the President of Taizhou hospital for over 10 years, you work hard to improve the quality of hospital management and medical services. You spent a lot

L. Ge and Z. Wang, *Thoughts on Hospital Design and Construction in China*,
https://doi.org/10.1007/978-981-97-0818-5_2

Fig. 2.1 Haixiao Chen

of time to direct hospital design and improve the quality of hospital construction. Would you please talk about your experience connecting hospital construction and management?

Haixiao Chen: In my opinion, the quality of hospital management in China is at a relatively low level. In comparison with our peers in developed countries or other domestic industries such as the hotel industry, the concepts and methods of hospital management in China are less detailed and need refinement. Part of the management is related to the 'hardware' of hospitals, which is the built environment, including the buildings and constructions. However, hospital presidents need a better understanding of the 'hardware.' *It is because 99% of the hospital presidents in China are medical majors. Their knowledge of building design and construction is very limited.*

When we started working on a new building, I told myself: 'In fact, you don't understand (Fig. 2.2). Not only don't you understand but also you don't understand at all. You lack the knowledge of building and construction whereas you have the power to build.' *Lack of knowledge is normal. However, when it was combined with power, it would be a problem.* So, I remind myself from time to time: Be aware of what you can do and what you cannot. I can cure patients but I can't build buildings. However, I need to build buildings for our hospital, since me as the president can't trust others to do it. Buildings must be constructed for a hospital and they set up the platform for its development. A hospital president should firstly identify people who can build buildings and secondly strengthen self-learning to understand at least the basic and principal knowledge of building design and construction. In this regard, we lack an architectural knowledge training system for hospital presidents or managers who are in charge of hospital design and construction in China.

First of all, we should recognize our shortcomings. Secondly, we should be clear that design and construction professionals are required for hospital development. Thirdly, we should strengthen ourselves by learning relevant knowledge. It includes the state-of-art of modern hospitals, such as humanization, green architecture, the integration of medical service processes, and so on. To establish these concepts, we

Fig. 2.2 Taizhou Enze Medical Center (one of Enze Medical Group), Zhejiang Province

must have an overall goal that may be vague at the beginning but we can find ways to clarify it. Once we figure out the direction and the required tasks to achieve the goal, we can then find appropriate professionals to complete the tasks. In this regard, I deeply feel it is difficult and requires robust ability to integrate the ideas from our staff members and those from the experts. These ideas are often 'fragmented' and need to be integrated. *'Fragmented' knowledge or ideas are often correct by themselves but may not be applicable if mixed in a 'big machine.'* It really takes a lot of effort to integrate them.

Lun Ge: Your thoughts answer the question inquiring the vision and positioning that a hospital president should have regarding hospital design and construction. I hope all hospital presidents can borrow your thoughts. Improving one's vision is the first step when starting a hospital building project.

Haixiao Chen: Therefore, *as hospital presidents, we should view building design and construction from a relatively macro perspective in order to better control the direction of development and set up requirements by level.* Even if a hospital would like to generously invest in construction, the president still needs to refine construction tasks in order to ensure that specific hospital needs are appropriately addressed. *If I were the president, I wouldn't go for an expensive building. I may not be able to build the best building but a building that suits our hospital the best.* It should meet the needs for medical services and reserve spaces for a five-to-ten-year development. Reserved spaces should be saved "blank" and provide opportunities for future managers to pursue their deeds. In our hospital, the land for the 2nd and 3rd phases of development has been well saved. It isn't necessary for future presidents to tear down existing buildings in order to pursue future development. Development can be smoothly conducted along with the existing buildings.

Lun Ge: Your considerations and activities are valuable, especially in the context of China's fast growth. Future presidents must appreciate. My second question is regarding the renovation and expansion of Taizhou Linhai Hospital. What's your experience renovating and expanding this hospital?

Haixiao Chen: *Regrading hospital renovation and expansion, we should firstly consider to save and reach the full potential of existing buildings. We should respect every piece of brick.* You may not like the existing but it should be treated with respect. For instance, a tree in front of our hospital has been there for 30+ years. We won't tear it down as long as it doesn't significantly influence our services. As far as we can arrange our work in harmony with the existing buildings, we do not need to build a new one. This is our philosophy. We value hospital planning since renovation and expansion are all based on the existing situation.

In my opinion, one issue of the hospital design and construction in China is that everyone values hospital planning but no one knows how to plan. Many hospitals criticize others for not having a plan whereas they have the same problem unsolved. It would be lucky for them if not being criticized in the next three or four years. It would never be too late to work on hospital planning, especially when the overall design was weak.

On the other hand, I think we should do a 'rich planning', that is to say, we must spend a lot of time, money, energy to develop the overall plan. Although we have to budget hospital construction, we need to enrich the planning. Even if we already had a well-developed or matured plan, we would need to re-think about it in order to identify and correct any imperfections. In early planning, we must ensure the fundamental elements are well organized, including functional spaces, service efficiency, and so on.

Therefore, the competitiveness and quality of a hospital building are influenced by two factors: overall planning and building design. It would be better if spending time on these factors during the period of planning other than reconstructing to solve the problems found after building occupancy. *The design of a hospital building should be in harmony with not only the environments inside the hospital but also the surroundings.* Regarding the renovation of our hospital, I suggested to 'replace the original with the original.' The style and fundamental function of a renovated building should be similar to its original and with no 'outstanding' looking or significant functional changes, which would increase the budget of renovation. So far, the renovated Enze Memorial Center in our hospital goes very well with its surroundings.

Lun Ge: We visited Enze Medical Center this morning. The construction of a new building and the expansion of an existing building may be different, in terms of project planning and positioning. Please talk about your experience constructing a new hospital building.

Haixiao Chen: Regarding a new hospital building, it is important to have a vision of development. Compared with renovations, developing a new building may be relatively easy for the president. *Decisions on a new hospital building should be consistent with the overall planning, goals, and value systems.* It is related to patient-centered services, high quality of care, science and technology supported development, etc.

There are a lot of contradictions regarding the development of a new hospital building. In order to reduce the contradictions, we should firstly respect every factor involved in the development: patient needs, hospital management, design concept, building codes and standards. Secondly, *we should value hospital users' empirical experience. Highly idealized architecture may not provide good user experience.* The stronger the theory, the more likely bad user experiences. *In my opinion, emphasizing the importance of user experience is the core of hospital design.* During the tour of our new building, you suggested wooden handrails since the stainless-steel material currently applied on the handrails felt cold. When we talk about humanization, I think it means warmness. *Humanization design of a hospital should help to create warm atmosphere in the hospital. Compared to the theory of humanization, the experience of room temperature is more intuitive.*

Lun Ge: That is a very good point. Room temperature can be measured so that humanized design can be quantified. You have a lot of opportunities to communicate with architects. Would you please give them advice?

Haixiao Chen: *I hope architects can view design as an art work rather than a business product. Architects should balance customer needs, designer's preferences, and styles. Based on the balance, a synthesis of values can be built.*

Lun Ge: It should be an architect's fundamental goal for career.

Haixiao Chen: *The difficulty lies in the creation of an organic, coordinated, acceptable, and well-experienced complex.* Secondly, I hope to see architects continuously improving the quality of their design. High-quality design can't be completed overnight and needs constant improvement. I hope architects can constantly improve the quality of their design during the phases of building construction and occupancy. *In this respect, I think architects should design from the standpoint of building users.* Users of a hospital building are patients and hospital staff members, whereas the hospital president may in charge of it during the phase of building design and construction. Hospital *design should respond to the needs of patients and staff members.* The ultimate goal of a hospital design is to optimize the relationship between the building and its users, including patients and the staff.

Lun Ge: Your thoughts hit the essential point of hospital architecture: who are the users. At present, many hospitals are built in response to the ideas of local governments or hospital presidents, overlooking the needs of patients and staff members.

Haixiao Chen: *A hospital president should serve as a representative of the patients and hospital staff, a bridge between building users and architects.*

Lun Ge: The new building of Enze Medical Center Fig. 2.2 was designed by Japanese architects. Would you please talk about your experience working with Japanese architects?

Haixiao Chen: The reason we choose a Japanese design team was their efficient use of resources in design. Hospitals in China shouldn't compete for being big in size or blindly admire the design of hospitals in developed countries. Public hospitals in China are welfare organizations and thus require an efficient utilization of resources. In this respect, the characteristics of a Japanese hospital design meet our needs. So, we chose a Japanese design team and so far, we have been satisfied with their work. *Compared with other designers, they can better control precise design and design details.* They regularly visit construction sites and follow our agreement on design. They conduct good negotiation with us while adhering to their design ideologies, such as the decisions on building colors and materials. Meanwhile, we value their practical experience designing medical buildings. They have developed not only a set of widely-recognized hospital design index, but also a set of design data collected by themselves, which are according to years of practice. In general, we choose a Japanese design team because our value systems on design are similar; secondly, they have rich practical experiences; and thirdly, they have a dedicated and rigorous attitude on hospital design.

Lun Ge: Ten years ago, I introduced the concept and methods of pre-design and planning for hospital projects, which should be the earliest introduction on hospital design research in China. The pre-design and planning of your hospital were initiated during that time (Fig. 2.3). Please share your opinions in this regard.

Haixiao Chen: In 2005, I learned that Professor Lun Ge in Beijing led an NSF research project on hospital design and also practiced in hospital pre-design and planning. So, we invited you to join us.

Lun Ge: It was our honor to join you. It has been almost a decade since our collaboration on the pre-design and planning We can tell that you have realized the importance of pre-design and planning since then. Would you please share your experiences? Any comments regarding the pre-design and planning we have done for your hospital?

Haixiao Chen: Pre-design and planning should be done before schematic design. *Many hospitals know the concept of pre-design and planning but don't know how to do it.* At that time, you were also groping for a complete process of planning, including process analysis, chart analysis, and so on. The planning has been very helpful not only to the prior hospital but also to our new hospital. *I think the pre-design and planning help hospitals to clarify what they need.*

Lun Ge: An important aim of pre-design and planning is to bridge architects and hospital managers. It is also a process of sharing thoughts and ideas.

Haixiao Chen: Yes, *when hospital managers cannot clearly express their needs* by using *architectural language, pre-design and planning bridge the thoughts.* The

Fig. 2.3 Taizhou Hospital (one of the Enze Medical Group), Zhejiang Province

method of pre-design and planning can be generalized in many fields. It is an indispensable link to improve the quality and efficiency of design.

Lun Ge: We appreciate this opportunity to collaborate with you. We get more matured through practice. The last question, and also the most important one is: at present, many hospitals focus on marketing and thus invest a lot in the decoration of building facades and interiors to attract customers. As a hospital president, what do you think about this phenomenon?

Haixiao Chen: As what I just said, every developer has his or her own pursuit or sense of achievement. We want to build the most suitable buildings for our hospital. It can be moderately advanced but must stand on solid ground. We may raise money for hospital development in many ways, but we can't endlessly build in debt. There should be a set of standards to define and evaluate hospital design and construction. *There is a five-star evaluation system for hotels. I look forward to seeing similar standards for hospitals.* The standards shouldn't only be based on hospital size, but also based on the extent to which a hospital completes its services. I think hospitals

Fig. 2.4 Haixiao Chen and Lun Ge

can borrow construction experiences from hotels. Most public hospitals in China should be set between two and three stars. We may build some four-star hospitals. Five-star standards are only suitable for some specialized or private hospitals. In this regard, it is necessary to learn from the industry of hotels. *We should reserve the best built environments and equipment for patients, just like hotels serve the best for their guests. Our work should be patient-centered and focused on patient and staff satisfaction.*

Lun Ge: Then I would like to extend the question. From the perspective of a manager, what do you think of patient-centered services?

Haixiao Chen: Give patient care the most priority in hospital services. A hospital should organize the best resources for patients, in terms of space using, medical equipment, transportation, in order to promote convenient, quick and warm services.

Lun Ge: That is a wonderful interview. Thank you very much (Fig. 2.4).

2.2 Design a Hospital with Soul—An Exclusive Interview with Jie Qiu

Time: January 2015—reiterated in May 2023
Location: Gansu Provincial Maternity and Child-care Hospital, Lanzhou

Jie Qiu: President of Gansu Provincial Maternity and Child-care Hospital, Chief Physician, Professor, PhD Supervisor; 1st-level Leading Talents of Gansu Province,

Fig. 2.5 Jie Qiu

Awardee of the National Labor Medal and the Excellent Hospital President in China (Fig. 2.5).

Key sentences:

1. *Today's hospital development shouldn't be focused on the piling up of expensive building materials, but help people to get the feeling of returning to nature and having the inner harmony with nature, leading to stimulated self-healing.*
2. *The first task is to improve the quality of populations. The second is to improve the quality of life. Therefore, the "soul" of our hospital is to contribute to these responsibilities and duties entrusted by our people.*
3. *I hope that forward-looking building standards and norms for maternal and child care can be well developed in line with China's reality. The standards should be oriented and also a support point for promoting the quality of populations. This is by no means building a maternal and child-care hospital can solve all the problems, but the responsibility should be carried out and promoted by the hospital.*

Lun Ge: Since you were selected as the President of Gansu Maternity and Child-care Hospital, both the enthusiasm of physicians and staff members, and the social and economic benefits to the hospital have been increased. It has been two and a half years since the new hospital building was in use. In the light of today's achievement, would you talk about the goals put on to the new hospital building at the beginning of construction? And how much have the goals been achieved?

Jie Qiu: It took eight years for us to complete the project (Fig. 2.6). From nowhere, we explored and harvested all along the way of design and construction. *Instead of viewing a hospital building as a solidified work of art by the hospital and its designers, I think it is a spatial construction endowed with life and soul. In a high-quality hospital, people experience not only the appearances of spaces inside and outside the building, but also humanity communication, sublimation, continuity, and the development of life.*

Fig. 2.6 Gansu Province Maternal and Children's Health Center, Gansu Province

At the beginning of a hospital project, the development team should be clear about the "soul" of the hospital. If this can be clarified at the early stage of design, the team will grow well during the project. The hospital should systematically think about its ideological system, value orientation, responsibility, and connotation of culture. Based on this, architects may express the essence of hospital culture in architectural language. If the hospital itself didn't have a deep understanding and a clear statement of its needs, designers and architects wouldn't be able to draw up an ideal blueprint.

During the preparation of our hospital, we conducted a literature review and surveyed physicians and staff members to investigate their needs. The focus was on their suggestions regarding room sizes and functions, in order to keep pace with the state-of-art of hospital development. Later, after exchanging design ideas with you, we developed a better understanding of healthcare architecture and clarified what we should do for the new hospital. We were thinking of a 3-in-1 hospital. It should be an environment providing healthcare services, a museum of medical science, and an art museum. The original idea was to change the traditional perception of hospitals, emphasize the benefit of hospital services on mental health, and "copy" or borrow some western concepts of modern architecture.

Therefore, when you worked with us on the research regarding spatial design for medical services, processes, and technologies, we doubted it: "Do we really need this research?" Today, we are sure that the research for preliminary planning is necessary and a lot of efforts must be made at the level of preliminary planning and design. The hospital development team must have intensive communication with designers, integrate ideas during the process of design, and sublimate a theoretical framework of design, leading to a complete architectural work with life and spirit. I still remember the time when you and our team sketched up the first plan of our new hospital in a hotel room in Shanghai. The sketch should be framed as a souvenir. Innovated by you, the design concept of building two green yards inside the hospital have come true.

No design can be perfect. However, if the preliminary planning research can be done well, there would be less regret. The efforts we made at the early stage of design paid off and promoted humanistic care and work efficiency. For example, regarding the appropriate location of the toilet in a patient room, there were two options: near the room entrance or by the room window. During the early design, we collected data on staff's walking routes and the distance of their daily walking for patient care through on-site observation using pedometers. It was found that if the toilet were located near the entry door, staff need to cross a 2.5–3-m corridor in order to see the patient. Otherwise, the patient can be seen by the staff at the moment of opening the door. Thoughts and perceptions on these design factors are different.

Statistical analysis showed that if the toilet were located by the window, the average duration of patient-staff communication in gynecology wards was 1.3 times longer, and the increase in obstetrics wards was 80%. Because the by-window location supports toilet ventilation directly with the outside, the problem of toilet smell can be effectively solved. *An important concern in hospitals is how to increase the validation and frequency of patient-staff communication. The communication plays an irreplaceable role in improving the quality and safety of patient care. I think that promoting patient care by design is the real purpose of healthcare architecture.*

What is the core of health care in the future? To activate and stimulate human potential and self-healing. How to activate and stimulate it? It should rely on the harmonious coexistence of human being and nature. Therefore, today's hospital development shouldn't be focused on the piling up of expensive building materials, but help people to get the feeling of returning to nature and having the inner harmony with nature, leading to stimulated self-healing. During the design of a hospital, each detail should embody its culture, vision, and value systems. *Hospital design is not only the development or construction of a building, but actually a process of launching and transferring culture into civilization.*

In our case, the goal and vision of our new hospital building were set up at the beginning of project. We would like to see it as a competitive world-level maternal and child-care hospital in China. With this goal, the needs and functions of our hospital can be clearly defined and then move to the process of design on the spaces, functions, and technology. In the process of design, hospital culture is especially important. Building functions actually intend to present the hospital's "soul".

First of all, it is a maternal and child-care hospital. The core tasks of our services have been well defined by the national laws and norms. *The first task is to improve the quality of populations. The second is to improve the quality of life. Therefore, the "soul" of our hospital is to contribute to these responsibilities and duties entrusted by our people.* Building design and construction of a hospital should never be led away from its positioning. *If a hospital messed up on its strategic positioning and were directed by market benefits, say, promoting some popular disciplines for marketing, the hospital would be a "spirit–splitting" building without soul.*

Secondly, we should carefully analyze the cultural concept of a hospital and then express it in architectural language. For example, the aim of our service is to address the needs of our people over everything else. Therefore, during the construction of our building, we laid 180 rescue units on the first floor, installed large-scale German generator units on the second floor, and set up nearly 5000 m^2 of emergency spaces to be prepared for war. Why did we do this? Some people may think that these spaces are useless in reality and waste money. *However, if natural disaster or war occurred, hospital buildings shouldn't only transmit lights but also make a strong sense of social responsibility, which bring warmth and hope to people. This is the "soul" of a hospital building.*

Moreover, modern healthcare services should take patients' psychological needs into account. The highest realm of healthcare includes both physical and mental health. *However, current healthcare services are more concerned about physical health, overlooking the needs and demands for healthy spirit and soul. Today, when building a new hospital, we should work hard to bring the goal of healthcare services back to its original. From certain point of view, mental health is more important than physical health.*

In the design and construction of our new hospital, we developed the main 'street' or corridor linking varied functional spaces into a gallery of art work focusing on "humanity and love", to make this scientific medical environment as an art museum. *If today's hospitals can consider their social responsibilities and address the needs for social development, we would better address the overall gaps of healthcare quality between China and developed countries.*

After eight years of hard work, we finally completed the construction of our new hospital building. Throughout the entire process, we explored and learned, transforming our perception regarding a hospital building from a mere work of art into a living and breathing space. We realized that a high-quality hospital is not just about the physical appearance of the building, but also about the human experiences, communication, and the development of life within its walls.

From the outset of this hospital project, we recognized the importance of a good understanding regarding the "soul" of the hospital. By clarifying our ideological systems, values, responsibilities, and cultural connotations, we could effectively express the essence of our hospital's culture through architectural language. We conducted extensive research, including literature reviews and surveys of physicians and staff members, to identify their needs and incorporate the latest advancements in hospital development. This process allowed us to create a unique vision for our new hospital: a 3-in-1 facility that serves as a healthcare environment, a museum of medical science, and an art museum. Our goal was to challenge traditional perceptions of hospitals, prioritize mental health, and incorporate concepts from modern architecture.

During the initial design phase, we collaborated closely with you, the designers, and our team to develop a comprehensive understanding of healthcare architecture. I still remember the moment when we sketched the first plan of our new hospital in a hotel room in Shanghai. It was a significant milestone in our journey. With your innovative ideas, we were able to realize our vision, including the design concept of building two green yards within the hospital.

We quickly learned the importance of thorough preliminary planning and research. Our efforts during this stage greatly influenced the humanistic care and work efficiency within the hospital. For instance, we meticulously considered the placement of toilets in patient rooms, conducting on-site observations and statistical analyses to determine the optimal location. By placing the toilets near the room windows, we observed a substantial increase in patient-staff communication and solved the problem of odors. Such design considerations directly impact the quality and safety of patient care, underscoring the true purpose of healthcare architecture.

Looking towards the future, we believe that activating and stimulating human potential and self-healing is at the core of healthcare. To achieve this, the harmonious coexistence of humans and nature is crucial. Therefore, our focus shouldn't be on extravagant building materials but on providing an environment that fosters a connection with nature, leading to stimulated self-healing. Each detail in the hospital's design should embody our culture, vision, and value systems. *Hospital design is not merely about constructing a building, but rather a process of launching culture and transferring it into civilization.*

Our goal for the new hospital building from the beginning of the project was to become a world-class maternal and child-care hospital in China. This goal allowed us to clearly define the needs and functions of our hospital, guiding the design process for spaces, functions, and technology. *Throughout the design phase, we remained mindful of our hospital's culture, ensuring that the building's functions would reflect our hospital's "soul."*

As a maternal and child-care hospital, our hospital has the core responsibilities to improve the quality of populations and enhance the quality of life. We placed a strong emphasis on meeting these obligations entrusted to us by the people. Our construction choices, such as installing rescue units, generator units, and emergency spaces, reflect our sense of social responsibility. While some may question the practicality of these spaces, they serve as a symbol of hope and warmth during times of natural disasters or wars.

Furthermore, we recognized the significance of addressing patients' psychological needs alongside their physical well-being. Mental health is as important, if not more important, than physical health. To promote holistic healthcare, we transformed the main corridor of our hospital into an art gallery that focuses on "humanity and love." This artistic approach aimed to provide a healing environment while fulfilling our social responsibilities and addressing the overall gaps in healthcare quality between China and developed countries.

In conclusion, our journey to construct the new hospital building was a transformative experience. We shifted our perspective from a static work of art to a dynamic space that promotes life, communication, and human potential. Thorough preliminary planning and research were vital in achieving our goals and fostering humanistic care. By staying true to our hospital's vision, values, and cultural concept, we are able to create a hospital that not only serves its service purpose but also serves as a beacon of hope and compassion for our community.

Lun Ge: During our collaboration on research for preliminary planning, we were deeply moved by your passion for patient care. We understood your passion and goals set up for this project. Can you talk about the special work done by your team?

Jie Qiu: When it came to building decoration, our considerations extended beyond budget constraints to meet the spiritual needs of our patients. We raised the average floor height to create spacious interior spaces that reduce stress and promote a sense of openness. In common areas and corridors, we carefully selected sound-absorbing materials for the ceilings. These materials are cost-effective and offer excellent sound-absorbing effects, fostering a natural sense of tranquility and peace of mind. Furthermore, we adorned the walls with paintings and artwork that revolve around the theme of "humanity and love," aiming to evoke warmth and provide solace to those within our hospital. We firmly believe that healthcare environments should instill a sense of calm and peace. This attention to detail in building decoration is an integral part of patient care and plays an irreplaceable role in soothing and healing the heart.

When it comes to the utilization of our building, we strive to honor human needs and the authenticity of medical science. Within our premises, we have incorporated an academic auditorium that serves multiple functions, including simultaneous interpretation and video broadcasting. This space is open to the public and hosts free healthcare educational classes and welfare films. The aim is to disseminate healthcare knowledge and share outstanding works of art. In the lobby of our inpatient department, we have even installed a high-end grand piano, typically found only in concert halls. Equipped with LED video screens, this piano serves as a source of daily classical music played by volunteers from local universities for the enjoyment of patients and staff members. The presence of music allows everyone in the vicinity to experience the allure and warmth of art, fulfilling our noble aspirations and goals in healthcare. Therefore, a hospital should not be a sterile medical environment but a celebration of human nature and life. Love should be the defining charm of a hospital.

In summary, our approach to building decoration takes into account both the financial considerations and the spiritual care needs of our patients. By creating spacious and serene interior spaces, using sound-absorbing materials, and incorporating artwork that embodies the theme of "humanity and love," we aim to provide a comforting and soothing environment. Additionally, our focus on building utilization involves the establishment of an academic auditorium for educational purposes and the inclusion of a grand piano in the lobby to offer the enchantment of music. Through these thoughtful design choices, we strive to elevate the healthcare experience, paying homage to the intrinsic value of human life and promoting an atmosphere of love within our hospital.

Lun Ge: The contents and feelings you described are exactly what some hospital buildings lack. Compared with other building types, the design and construction of a hospital would be more likely to initiate its culture. Would you please talk about any regrets regarding this project?

Jie Qiu: The construction of our new hospital building was a crucial part of the expansion of our existing hospital, but we faced significant challenges due to limited land and space availability (Fig. 2.7). As a result, many original design ideas and proposals, such as a solarium, activity rooms, training rooms for early childhood development, rehabilitation areas, and spaces for new disciplines, couldn't be realized. Despite this constraint, the demand for healthcare services quickly filled the 1000 patient beds in the new building, with the number of patients continuing to rise. It is evident that the shortage of healthcare spaces and the increasing number of individuals in need of care have become pressing issues in the field of maternal and child care.

Fig. 2.7 Aerial view of Gansu Province Maternal and Children's Health Center, Gansu Province

To address these challenges effectively, it is essential to seek national-level support and the development of policies specifically focused on maternal and child care. Updated standards for scientific and systematic development and construction are also crucial. Furthermore, it is important to recognize that there is still a gap between hospital development in China and that of developed countries. This gap extends beyond building design and material usage and encompasses a disparity in understanding and cognition of healthcare services.

As the saying goes, "Practice without theoretical guidance is a blind practice." Therefore, it is crucial to establish forward-looking building standards and norms for maternal and child care that are in line with the realities of China. These standards shouldn't only be oriented toward improvement but also serve as a support point for enhancing the quality of populations. It is important to acknowledge that the construction of a maternal and child-care hospital alone cannot solve all the problems, but the responsibility for promoting the quality of populations should be shouldered and advanced by the hospital.

In summary, the limitations in land and space availability during the construction of our new hospital building have posed challenges in realizing certain design ideas and meeting the growing demand for healthcare services. To overcome these obstacles, it is crucial to garner national-level support through policies and updated standards that align with China's reality. *Furthermore, bridging the gap between China's healthcare development and that of developed countries requires a good understanding of the comprehensive nature of healthcare services.* By actively addressing these issues, we can work towards promoting the quality of populations and fulfilling the responsibilities of a maternal and child-care hospital.

Lun Ge: Any expectation for medical architects?

Jie Qiu: Chinese architects often face a common challenge rooted in their cultural background. Influenced by traditional values, many Chinese architects tend to exhibit a relatively introverted sense of design and could benefit from adopting a more outward-looking approach. *A design proposal should demonstrate architects' deep understanding of the project's orientation and functions, leading to its elevation to a higher level.* Unfortunately, some designers may limit themselves to merely replicating fashionable concepts or being driven solely by market trends and economic benefits.

However, it is important to note that the architectural landscape is evolving, and we are witnessing positive changes among Chinese architects today. There is a promising outlook for the future, as we hope that our architects and designers will continue to embrace new ideas and evolve their design conceptions. By expanding their horizons and staying updated on emerging trends, Chinese architects can unlock their full potential and contribute to innovative and transformative architectural projects.

Lun Ge: Thank you very much (Fig. 2.8).

Fig. 2.8 Jie Qiu and Lun Ge

2.3 Making Functional Master Plans—An Exclusive Interview with YunXiang Liu

Time: March 2015—reiterated in May 2023
Location: Huwan Hotel, Beijing

YunXiang Liu: Former Party Secretary and Deputy Director of Health Bureau Yantai, Shandong. Former Party Secretary and President of Yuhuangding Hospital, Yantai. Holder of the honorary titles "National Model Worker", "Shandong Province Model Worker", "National Outstanding President." Awardee of the "National Outstanding Contribution Award in Hospital Management" and the "National Hospital Construction Lifetime Achievement Award." (Fig. 2.9).

Fig. 2.9 YunXiang Liu

Key sentences:

1. *Regarding hospital management, I think the key is having a theory of development and a good understanding of the functional flows and their relationship.*
2. *Hospital managers shouldn't rely on architects and planners to design a dream hospital since these designers aren't medical professionals. Designers typically design based on a design concept. As medical professionals, hospital managers should be involved in the design of medical service processes, department functions, and design details.*
3. *The key for hospital planning, design, and management is to promote the process of medical services. Integrated planning and humanization design from the standpoint of patients are preferred.*

Lun Ge: Hello, Director Liu. The new Yuhuangding Hospital applies a new concept of hospital design and planning. Nationwide, it is the first hospital project proposing patient-oriented services. There is a trend called "Learning from Yuhuangding Hospital." As the leader of this project, what do you think about this phenomenon and this project? Can you introduce the process of decision making during the period of planning and design?

YunXiang Liu: The previous Yuhuangding Hospital was built in 1890 and developed as a Church hospital by an American Presbyterian Church. Its appearance was in a traditional European style. Our team decided to keep it in the new hospital. We also decided to add modern elements to the new. *The new hospital was built with our respect to the hundred-year history, accumulating the traditional and modern cultures.*

At the beginning of this project, our team collected suggestions and conducted consultations regarding hospital planning and design (Fig. 2.10). However, no systemic theory was developed. We had a common idea of revising the traditional appearance. It was our initial consideration regarding the design. We decided to use creative design ideas to guide the project development. Patient-centered care is the key. *From the standpoint of a patient, I keep thinking of how to build a patient-centered hospital. One of the ways is to put an emphasis on humanization in design.*

Fig. 2.10 Yuhuangding Hospital, Shandong Province

In hospital design, humanization can be addressed in three aspects: functional flows, services, and the environment. For patients, humanization can be promoted by optimizing their visiting experiences and maximizing their convenience in the hospital. *The new Yuhuangding Hospital was built in 2001 and made many national records.* For example, it was the first hospital using stratified registration systems for patient check-in/ out and medicine *pick-up. These new systems provide convenience to patients and reduce their stress associated with unnecessary travels between doctors' offices and laboratories. Moreover, the IV (intravenous)* units for outpatients and emergency patients are combined for their convenience. Laboratory resources are also arranged for sharing among multiple departments to optimize services. These design factors are based on humanization and intent to optimize services.

Regarding the environmental design and spatial functions of a hospital, I would like to share some information in detail. At the time when this hospital was built, the national standards for inpatient units suggested a width of 2.4 m for corridors. However, our team set up higher standards and the average width of corridors in our built hospital is 2.8 m. Moreover, waiting areas in this project are designed to be open spaces with high ceilings to promote environmental and psychological comfort for patients. In the pediatric department and obstetric department, we widened the width of a patient bed into 1.1 m instead of the typical width of 0.9 m, in order to meet families' needs for accompanying children.

In the pediatric department, we specially designed and installed dual-use toilets for both adults and children to promote convenience. To further enhance humanization, we leveled up the standards for patient room design in terms of room width and depth. In the first-phase of this project, the average width of the column grids in inpatient buildings was 7.8 m. It was increased to be 8.2 m in the second and third phases. The new patient rooms were thus spacious enough to comfortably accommodate three beds in one room.

Because of the increased width, multi-function chairs can be included in patient rooms for family members. These chairs can be used as guest chairs during the daytime and unfolded to be accompanying beds for family members' overnight stay. In addition, we included a flower-stand in every patient room to improve its visual quality. *In one word, humanization in a hospital design is beyond the design itself and should be led by a good understanding of patient care.*

Lun Ge: It is surprising to know that the concept of humanization design has been applied in your hospital since 2001. Yuhuangding Hospital has been under constant renovation and expansion for more than 10 years. Are there any consistent guidelines for the planning, say, from the beginning to present, and for the future? Have these guidelines been adjusted to adapt to changes? What are the experiences and lessons learned during the growth of your hospital? Do you mind to share your thoughts regarding the relationship between hospital management and building development?

YunXiang Liu: In our hospital, building updates and retrofitting have been done consistently since 1998. We always use the theory of integrate design for hospital development. Overall, we must have a good understanding of the entire project.

Firstly, the "foundation" should be well developed. We should consider the development of a building from multiple perspectives, such as what kind of foundation should be set up and how it can be connected to possible extensions in the future.

In the first-phase of our new hospital project, 57,000 m^2 were built in the outpatient complex. With 9 floors, this complex has a standard floor plan in a shape of '日'. The first 5 floors are designed for the outpatient department. The inpatient department is located on the 6th to 9th floors. At each of these floors, three separated ward-districts are well connected to facilitate services. On the 6th floor, there are two roof gardens with natural lighting and ventilation. By doing these, we promote the efficiency of daily operation and the utilization of spaces.

The new complex was open in 2001. So, what should we do with the prior inpatient complex? What should we build if we pull it down? There were a lot of voices and thoughts. *I insisted on the principle of "integration". We must ensure horizontal circulations on each floor so that we can reduce the pressure on vertical circulations and integrate the spaces.* By doing these, we smooth the connections among the buildings, which are critical to hospital services. Meanwhile, we saved our original thought of locating relevant departments on the same floor for service efficiency.

The integration of buildings became the base of our project planning and enhanced sustainability later on. Horizontal circulations successfully facilitated services in the new buildings and thus improved the functionality of the entire hospital. Separated departments and buildings are linked to each other and work as one system. I would like to emphasize: *"the foundation" of a project must be well developed to ensure its growth.*

Regarding hospital management, I think the key is having a theory of development and a good understanding of the functional flows and their relationship. It is very important for hospital managers to establish this understanding, which can benefit their collaboration with designers. For example, hospital managers should be able to read architectural drawings and familiar with the varied functional flows in a hospital and the specific functions of each department. Once such knowledge was established, they would have the ability to provide appropriate suggestions on design. "*No excellent work can be developed without a pinky client.*"

Hospital managers shouldn't rely on architects and planners to design a dream hospital since these designers aren't medical professionals. Designers typically design based on a design concept. *As medical professionals, hospital managers should be involved in the design of medical service processes, department functions, and design details. They should follow up the design and eventually help to ensure the construction.* Therefore, I don't suggest the new systems titled "Contractor or Agency Construction" for hospital projects. Moreover, hospital managers shouldn't rely on supervising companies to ensure the quality of hospital building construction. *Hospitals should build their own teams to supervise the design and construction. It is the responsibility of every hospital and we should implement it in our daily work.*

Importantly, the overall concept of planning influences hospital development. *We need to consider the use of land and the reservation of land for future extensions. Different phases of the planning should be addressed as an integrated project.* We need to think about the distances between buildings, logistics, and the management.

These factors should be considered as one group. Actually, if the functional flows in a hospital were appropriately arranged, these factors can be smoothly addressed. Moreover, the process of hospital construction is logical. As long as we concentrate on the key factors including the flows, functions, and the environments, successful hospital planning, design and construction can be achieved.

For instance, in order to optimize the process of medical services, we have applied the concept of large departments since 2001. This concept is internationally advanced and years ahead of the similar concepts applied in other countries. For instance, there are 7 ward districts on each of the floors. It is reasonable to arrange the Neurology Department with 2 of the districts, the Neurosurgery Department with 3, and the Orthopedic Department with 2. By doing this, group consolations, rescues, and transfers are promoted. On the other hand, the Gastroenterology Department, the Digestive Surgery Department, the Cardiovascular Medicine Department, and the Cardiovascular Surgery Department are located on the same floor. *The new Yuhuangding hospital was the first Chinese hospital built with the concept of large departments.* Nowadays, this concept and relevant studies have been introduced domestically and internationally. Arranging relevant departments on the same floor can be a good benchmark for new hospital design (Fig. 2.11).

The key for hospital planning, design, and management is to promote the process of medical services. Integrated planning and humanization design from the standpoint of patients are preferred. By doing these, we can easily handle the problems frequently reported on hospital construction and management.

Lun Ge: President Liu, your experience is quite profound and useful. You mentioned the needs for high-quality hospitals. How to make it happen?

Fig. 2.11 Aerial view of Yuhuangding Hospital, Shandong Province

YunXiang Liu: *A firm belief of patient care and advanced ideas on design are required. Hospital managers need to be serious and persistent, and work hard to pursue a perfection.* They should observe things around them and learn from others. In 1992, right before the new Yuhuangding hospital project, I visited several hospitals in Australia and observed the processes of their services. I then considered how to modify our service processes, optimize the functions, and improve the quality of our hospital. I think about how to develop healing environments and help patients to feel comfort both physiologically and psychologically.

We should pay more attention to the details in a hospital. Details typically determine the quality. For instance, it is critical to promote natural ventilation and daylighting in patient rooms. *No room in our hospital is designed without natural ventilation. I strongly oppose this design of "black" rooms or rooms without windows.* All the rooms in our hospital have natural lighting and ventilation. *Moreover, we insist on including a window at the end of every corridor in order to improve mental and environmental comfort. Further, all baseboards along the corridors are arc-shaped to facilitate daily cleaning and etc.*

We pursued human-oriented and high-quality services during the whole process of building design and construction. Our persistence made today's Yuhuangding Hospital.

Lun Ge: Director Liu, what you have talked about is valuable. This project keeps pace with the state-of-art of healthcare in terms of planning, detail, and management. Mr. Lirong Zhuge, the Chairman of Hospital Architecture System Research Branch, said "The concept of Yuhuangding Hospital advanced the hospital design in China for 20 years". It is really beneficial talking with you. Really appreciate your sharing (Fig. 2.12).

Fig. 2.12 YunXiang Liu and Lun Ge

2.4 Persistence on Original Intents—An Exclusive Interview with Roberta Lipson

Time: March 2015—reiterated in May 2023
Location: Beijing United Family Hospital in Jiangtailu

Roberta Lipson: Founder and Vice Chair, United Family Healthcare, New Frontier Health (Fig. 2.13).

As a private healthcare organization, United Family Healthcare (UFH)'s success in China has not been easy. Our aim was never to make quick profits, but to create a premium, personalized, service-oriented platform for those seeking for humanity care.

Based on advanced medical concepts and practical experiences from the West, in the context of China, following the laws and regulations, we choose to balance ideal concepts and the reality to provide rational and innovative management. The goal is to provide safe, comfortable and premium-quality care to patients.

China's healthcare industry is still in transition. We hope UFH can present a new mode and concepts of healthcare. Together with the refinement of policies and regulations, we are committed to constant improvement of healthcare safety and service standards. The aim is to make high-quality healthcare services a priority. Meanwhile, UFH hopes to provide new standards and modes of healthcare for your reference.

Key sentences:

1. *Our mission has always been first and foremost to deliver the best medical care to our patients, and we believe that if we achieve that profits follow.*

Fig. 2.13 Roberta Lipson

2. *The core of high-end healthcare service is respect for patients in every way. It is a type of spirit and culture that come from the inside and must be internalized by every employee.*
3. *China's healthcare industry is still in transition. We hope UFH can present a new mode and concepts of healthcare. Together with the refinement of policies and regulations, we are committed to constant improvement of healthcare safety and service standards.*

Lun Ge: As a medical management expert from America, you have guided the development and operation of United Family Healthcare (UFH) in China for 25 years. The first UFH in China was built in 1997. Today, there are more than 11 UFH hospitals in province capital cities and many free-standing outpatient clinics. You must have a deep and multi-perspective understanding of Chinese healthcare systems, marketing, management, and operation. Do you mind to share your experience with us?

Roberta Lipson: First, the market for private premium healthcare is growing in China. The number of people who can afford to choose their own healthcare has increased substantially over the last 20 years. More people at the middle- and affluent-income levels are interested in the advanced life styles, improved services, reliability and convenience that premium private healthcare can bring. The aging population has increased financial burden on the public system. The size of the private healthcare market has been estimated to be over 90 billion USD.

From the beginning, our goal was to bring choice to the market. To be informed by the best practices globally, but to be culturally appropriate and compliant to the local situation.

Reasons for the successful development of UFH in China are as followings: First, we strictly abide by Chinese laws and the development of high-quality private health care are completely in line with Chinese policies. Meanwhile, we ensure the quality of our services meets the requirement of international standards. We are informed by evidence-based healthcare theories and focused on healthcare practice. To be more specific, we try to develop a balanced and ideal practice in the context of China, following Chinese laws and policies as well as international standards of health care. While innovating, we seek a balance which is informed by both International and Chinese norms of healthcare. *We created a new mode of care for the first UFH in China'*. At that time, there was no concept of a warm and homelike environment for healthcare delivery in China. *We took both international standards and Chinese characteristics into account and innovated UFH healthcare environments* (Fig. 2.14).

Second, we ensure consistent collaboration with high-quality healthcare professionals. Our healthcare experts include both Chinese doctors and professionals from other countries. Currently our professional staff is comprised of about 70% Chinese doctors and 30% from other Western countries. The number of Chinese doctors keeps growing since the proportion of Chinese patients significantly increases. Our foreign doctors are from multiple countries, including the USA, Britain, Australia, and Germany. We have unified UFH standards to ensure effective communication among doctors from different countries and thus maintain effective operation. For

Fig. 2.14 Entrance of Beijing United Family Hospital

example, JCI is one of the standards of UFH and it serves as our baseline quality requirement.

You may have a question: why many Chinese doctors' behaviors and work attitudes working with UFH are different from their previous ones working with public hospitals. It is due to the new environments, salaries, and patient-oriented service principles of UFH. These factors contribute to improved work attitudes, leading to the success of our mode. *At the beginning, UFH was recognized as an expensive hospital with good services and elegant environments. Now, we have also earned broad based respect and recognition in terms of professional and integrated medical quality and services, which attract excellent experts to join UFH from the best hospitals in China and from around the world.*

Third, UFH has a certain and unified mode of business management. Although our hospitals and clinics are scattered in many cities in China, we manage them through one shared management platform for efficiency and standardization. For example, all patient insurance, reimbursement, hospital finance, marketing, and human resources are managed and supported by this platform. As we build new United Family Hospitals, we have no need to start designing our systems from scratch. Moreover, our training programs (including those for clinicians, administrators and managers) and system-wide electronic medical records, IT systems, appointment systems are also shared. Led by the UFH board of directors, our management team formulates and implements UFH strategies and development plans.

Lun Ge: You just mentioned an integrated platform of healthcare services. Would you please talk about its development?

Roberta Lipson: We constantly learn and make improvements. Although the needs of Chinese patients and healthcare markets keep changing, we uphold our standards of healthcare and the principles of evidence-based medicine. We ensure the safety of patients and provide convenient, thoughtful, and high-quality services. *As the leading provider of high-end healthcare service in China, UFH leads the trends of healthcare marketing in many aspects.* For example, UFH introduced integrated homelike wards to China. *UFH is also the first facility to provide Family Medicine doctors to all of our patients to partner with them in crafting their wellness, prevention and chronic disease management plans and considers this as the core of healthcare.*

We follow international standards of healthcare and our Family Medicine doctors are responsible for our patients' primary care management and disease prevention. However, we have found that many people in China think that seeing Family Medicine doctor or GP for disease prevention and healthcare is unnecessary and prefer to see a specialist such as a pediatrician, obstetrician, or gynecologist. In order to adapt to these demands, we have adjusted our practice and encourage patients to choose a doctor, such as a pediatrician, obstetrician, or gynecologist, as their general-healthcare doctors, as alternatives to choosing Family Medicine. A patient and his/her general-healthcare doctor develop a long-term relationship of healthcare consultation and thus the doctor has a comprehensive understanding of the patient, his or her family, thoughts, and work. This makes UFH healthcare consultation more customized. For example, if a 50-year-old patient had no mammography examination, her doctor has the responsibility to remind the patient of the importance of this examination. Development and maintenance of this one-to-one personalized services and long-term consultation benefit both patients and doctors.

In addition, we work hard to provide convenience to our patients. First, along with the general hospitals of UFH, we have built multiple satellite clinics inside and around cities to reduce the distance that patients need to travel to see a doctor. Second, unlike many hospitals in China, we shorten inpatients' hospital stays at UFH hospitals. We advocate minimally invasive and day surgery where appropriate, allow inpatients to go back at home at an appropriate time after surgery and then can also send our doctors and nurses to provide follow-up care at their homes, such as stitches removing and post-surgery examinations. This helps to reduce patients' financial burden, increases their comfort, and tends to speed recovery. Third, our rehabilitation hospitals provide professional and convenient services. Post-surgery patients as well as stroke, cancer, and cardiac patients sometimes need rehabilitation care and physical therapy for a longer period after surgery which is safely and conveniently available in our rehabilitation hospitals.

In most western countries, most cancer patients receive chemotherapy treatment in outpatient departments. This is considered as a more safe and convenient approach allowing patients to carry on their normal life. Some patients also have reduced immunity at this time and are vulnerable to cross-infection and thus better off hospital environments. Therefore, UFH provides the option of receiving cancer treatment in a club-like outpatient cancer center, including radiotherapy, chemotherapy, psychological and nutritional supports. This type of arrangement not only meets Chinese patients' demands for care, but also provides safe and convenient cancer care at a high level.

Our healthcare system is managed online. UFH hospitals, clinics, and rehabilitation centers are managed by a unified information network. For instance, the information of a patient who moves from Shanghai to Beijing is saved in this network and can be accessed by our doctors at any UFH location. No matter where the patient receives examinations, the medical records are available for review in any UFH institution. Now we have an established cooperation relationship with some of the best academic medical centers in the US and can conduct remote group diagnosis regarding complex diseases. Moreover, all of our staff meetings are video conferences to ensure high-quality services in all UFH hospitals across China.

Meanwhile, we have cultivated our own next generation medical talent. At UFH, both Family Medicine and Emergency departments provide three-year on-the-job training for postdoctoral students to cultivate the next generation of UFH clinical leadership, which meets United Family Culture and international standards.

Many of China's other private facilities tend to specialize in one field. As a high-end integrated general hospital, UFH has obvious advantages. If a hospital were only specialized in gynecology and obstetrics, there would be risks on emergency care, for example, in case of massive hemorrhage. Since pregnant women may also have other illnesses, a general hospital can provide holistic service and better address patient needs for both physical and mental healthcare. Patients have both physiological and psychological requirements. Our responsibility is to address both demands and offer patients a sense of security.

Lun Ge: These aspects are indeed important. Would you please talk about the challenges you have faced during your work at UFH?

Roberta Lipson: *Our challenges involve two aspects: management and policy.* The challenge of management is that our services should be in line with international standards. So far, UFH has eleven hospitals and dozens of clinics in China (Fig. 2.15). All of them should follow unified quality and service standards. However, the reality is more complicated since these hospitals and clinics are located in different cities, doctors are from different countries, and patients are from over 100 countries worldwide.

The other major challenge is from policies. Each of the local governments has its own interpretation of national policies that that can be different from each other. In addition, while in general supportive of the development of private healthcare, the reform of policies is still in the process of evolution. So, we face challenges when we develop a new hospital in a new location and need to adjust our proposals to follow local laws and regulations. The national government encourages private hospital development and private investment in healthcare industry. We expect to see more refined policies and regulations soon.

Another issue is the compatibility of health insurances. Premium private hospitals like UFH have not been covered by the general healthcare insurance system of China so our patients need to buy commercial insurances. Hopefully, our patients can receive some reimbursement from the general healthcare insurance system and in the future

Fig. 2.15 Beijing United Family Rehabilitation Hospital

pay the rest of their expenses through commercial insurances. We have already seen progress in this area in some Chinese cities.

Lun Ge: This is indeed a big challenge and UFH's success is not easy. So, in your opinion, what is the main concept that leads to success?

Roberta Lipson: The main idea is perseverance. *Our mission has always been first and foremost to deliver the best medical care to our patients, and we believe that if we achieve that profits follow.* I came to China in 1979 when the healthcare system in China was underdeveloped. After more than 30 years' development, now both the healthcare technology and equipment in public hospital in China are infinitely more advanced. China has also made great strides in life expectancy for its people. Nevertheless, I hope that Chinese people can also enjoy the highest level of healthcare services including comfort and convenience.

Although UFH services cover a limited group of people who can afford to pay for premium services, our United Foundation for China's Health provides free healthcare services to the poor, disabled children and people who are in need of care. With our effort, we hope to introduce new modes and concepts to promote Chinese healthcare industry, such as giving more respect to patients. In the 1980s, our peers in China frequently mentioned their admiration of hospitals in the west, including small issues such as the absence of disinfectant odor in the hallways. This inspired me to consider all of the ways in which we could improve the service environment.

My major purpose is to do something meaningful for the society. Although we only serve for a limited group of people, it is meaningful. More importantly, we present a new mode and concept to the industry in China. Although we began over 20 years ago mostly serving international patients, Chinese nationals now account for over 70% of our patient mix. Along with the increasing availability of commercial health insurance and improving government policies, more Chinese people will choose high-end healthcare services. *The core of high-end healthcare service is respect for patients in every way. It is a type of spirit and culture that come from the inside*

Fig. 2.16 Roberta Lipson and Lun Ge

and must be internalized by every employee. To be more specific, this is a kind of corporate culture that employees should have the confidence to handle challenges in order to address specific patient needs. This way of caring is from the inside and not a simple posting or slogan. Only in this way can we offer the most humane care.

Lun Ge: Indeed, hospitals are not only containers of advanced medical technology but also sincere emotion. I am so honored to talk with you today and thanks you for sharing! (Fig. 2.16).

2.5 Optimise the Distribution and Manipulation of Medical Resources—An Exclusive Interview with Yingkang Shi

Time: January 2015
Location: Chaoyang District, Beijing

Yingkang Shi: Professor; Committee member of the PhD programs on cardiovascular and thoracic surgery, hospital administration, and medical informatics; former President of the West China Hospital affiliated with Sichuan University (Fig. 2.17).

Fig. 2.17 Yingkang Shi

Key sentences:

1. *Our top task is to make effective use of the limited high-quality medical resources available to us. Another issue is the limited "accessibility" of the medical resources. Accessibility should be improved to promote the efficiency of medical services. It requires efficient utilization of devices and spaces."*
2. *As a matter of fact, I think a preliminary planning of the overall project and the considerations on user needs are of the most importance. Moreover, during a hospital planning, we should take the situation of disasters into account since hospitals shall by no means stop saving patients in case of disasters.*
3. *As time goes by, there will be less resources worldwide and thus the global situations will be more severe: higher GDP, population aging, incurable chronic and geriatric diseases, and increased medical expenses. These are issues worldwide.*

Lun Ge: In the field of hospital design and construction in China, the unique architectural philosophy applied to West China Hospital surpasses all other hospitals. You are the president and the first planner of this hospital. What were the goals and objectives driving you to make decisions?

Yingkang Shi: Sichuan is a province with large populations but limited medical resources of a high quality. Our hospital is the only one that is affiliated with a university supervised directly by the Ministry of Health (Fig. 2.18). Although our hospital has a long history and relatively advanced medical technology, due to the high population density and the inland location of Sichuan, it is still less developed in comparison with the hospitals in central and coastal cities, in terms of service providing, education, and scientific research resources. Medical resources in Sichuan are scarce whereas the demands for medical services are at a high level. Many patients with critical diseases choose to travel to Beijing for treatment. It is a waste of local medical resources.

In the 1990s, we were in lack of economic capacity to augment the limited medical resources. At that time, financial supports from the central government to western

Fig. 2.18 Entrance of West China Hospital affiliated with Sichuan University (2015), Sichuan Province

regions were limited. Moreover, none of the limited supports from the local government would be sent to our hospital due to its affiliation with the Ministry of Health. However, the financial supports sent to us by the Ministry were very limited due to the location of our hospital—out of the national capital. When we worked on the planning of our new hospital, we had to consider how to address the big demands by appropriately allocating and deploying the limited resources. Regarding medical reforms, we have a lot to do. *Hospital planning asks for an overall consideration of the allocating and deployment of medical resources at macro, medium, and micro levels.* Otherwise, it would be impossible to address high-level demands with severely limited resources.

Our hospital focuses on intensive care and miscellaneous or critical diseases. Our top task is to make effective use of the limited high-quality medical resources available to us. Another issue is the limited "accessibility" of the medical resources. Accessibility should be improved to promote the efficiency of medical services. It requires efficient utilization of devices and spaces. Nevertheless, the problem of "accessibility" is inevitable. For example, many patients need to travel a long distance to see doctor. China is in short of high-quality medical resources and these resources are typically concentrated in big cities. We consider the costs of operation, possible outputs, as well as economical use of hospitals.

Secondly, at that time, our hospital had to rely on self-raised funds. It was very hard to collect funds. Today, our hospital has more than 5000 patient beds and 95% of the expenses for growth was paid by ourselves.

Under such circumstances, we carefully pay attention to the efficiency and performance of investments. We concentrated on how to organize the service processes and how to make our services better fit patient needs, economically and conveniently. In my opinion, many hospitals are slightly extravagant in building design and thus waste resources. For instance, many unnecessarily wide corridors or high ceilings have been built.

Fig. 2.19 Aerial view of West China Hospital affiliated with Sichuan University *in 2012*, Sichuan Province

Thirdly, regarding building management, from pre-design preparation to design management, construction, and building occupancy, we mainly focused on the following aspects:

Our first building was built in 1993. Once it was completed, we realized that it looked the same as many other hospital buildings. It was built for convenience and by land availability. This phenomenon should be changed. A scientific and rational planning should be made upon future development (Fig. 2.19).

Secondly, at that time, many architectural design institutes in China had no teams focusing on hospital design yet. Typically, hospital presidents are medical majors and have no knowledge of building design or construction. Once we built the first building, we made a right decision to hire experienced designers specializing in hospital design from foreign countries to make an overall strategic plan for our hospital. Probably, our hospital is the first one in China to do so. Once the plan was developed, we followed it and built the proposed buildings step by step.

The selected design institute was professional. They had developed methods of organizing the medical service processes, circulations, and functional relationships among hospital departments. The analysis of user needs was profound and professional. Along with the overall planning, they provided analytical data, including the estimated growth rates of outpatients, inpatients, and surgeries in 20 years. The design institute made a plan for us, which was based on an estimation of a 30-year growth. This plan was prospective and allowed flexibility. Since then, the quality of life in China have been significantly improved and our public healthcare systems have been reformed. These stimulated the needs for medical care. In about 10 years since the plan was developed, we have used all the building resources scheduled for a 30-year growth. For the time being, we have completed the construction of all proposed buildings. The only change we made on the overall plan was the addition of a parking lot and corridors. An open land of 30 acres in the center of our hospital was reserved for greening. Underneath the green area, we added a parking lot in the size of 40,000 m^2. We also added corridors between buildings to facilitate connections, which were overlooked in prior planning.

Thirdly, upon the completion of building construction, problems arouse gradually during the actual use of the buildings. For instance, the needs for vertical transportation were under-estimated in planning. The designer proposed the plan according to the references for American populations instead of Chinese populations. The needs of Chinese families for accompany inpatients were overlooked. Due to the financial stress in planning and our intendance of saving the best for patients, the spaces left for our physicians are very limited. This is a pity.

Throughout the planning, we adhered to the idea of conserving resources. Not only did we integrate the clinical laboratory and the radiology department which were traditionally separated from each other in hospitals, but also had we reorganized and adjusted the functional spaces in pharmacies. From the overall planning to detailed design and construction materials, we bore in mind at all the time the principles of resource conservation so that the cost of operation in our hospital is less than that of many other hospitals.

Moreover, I would like to share experience organizing logistical services. Our logistical services are planned with both our hospital and the surrounding areas in mind. The sterilizing room of our hospital has the capability to serve 10,000 patient beds, of which only about a half has been used by our hospital. The rest is available for other hospitals, such as private hospitals or other hospitals affiliated with the university. Likewise, the laundry room of our hospital also has the capability to provide services for other hospitals. In fact, extra building spaces and labors needed for the capability increase from 4000 to 10,000 beds are less than expected. The ideology of regional services has maximized the efficiency of resource utilization and facilitated administration. What's more, we built a central kitchen with a borrowed technique, 'rapid cooling', which is typically used for in-flight meals. At its full capability, the kitchen can provide about 40,000 meals per day, whereas our hospital needs about 25,000 meals or so. With better performance, it can provide services to kindergartens and schools, etc.

Lun Ge: It is of great practical significance. Both government policies on medical reform and hospital leaders should borrow your experiences optimizing resource use. Can you share more of your unforgettable experiences regarding hospital design and construction?

Yingkang Shi: *As a matter of fact, I think a preliminary planning of the overall project and the considerations on user needs are of the most importance. Moreover, during a hospital planning, we should take the situation of disasters into account since hospitals shall by no means stop saving patients in case of disasters.* There are three possible roles that a hospital may play during the time of a disaster: a saver, a victim in need for rescue, or both. To be a saver, the hospital itself should be safe and have the capability of saving people and providing medical care. In the second situation, the hospital itself is destroyed and in need for rescue. The third case means that the hospital is partly damaged but still capable of saving itself and offering help to others. Therefore, the decision makers should consider the various possibilities and find ways to be a survivor.

We have a central kitchen equipped with refrigerators in which raw materials, cooked or half cooked food are stored. In case of a disaster or war, our hospital can stand for 14 days without external supplies. If piped water sources were contaminated, we have the option to use water from deep wells that can provide 1300 tons of water per day to address the needs for medical services and surgeries. Moreover, on a daily base, each of our boilers has 10 tons of fuel to guarantee the steam power necessary for medical services. Further, each of our operating rooms and intensive care units is equipped with an electric generator. We also have a portable generator for emergency use. We have taken these things into consideration at the preliminary stage of hospital planning.

Furthermore, another issue in need of attention is the physician-nurse rate in China. The proportion of physicians is at a high level whereas that of nurses is at a very low level. Once we compared the rates between China and France. In a hospital affiliated with the University of Montpellier, 11% of the staff were physicians. Whereas, the percentage of physicians in our hospital was 22. This phenomenon means that we are in short of auxiliary staff members. For example, with the assistance from a secretary dealing with material collection, a doctor could spend 70% of his or her work time on important tasks for patient care, such as analyzing patient symptoms. In this way, human resources can be better organized. Meanwhile, if the physician-nurse rate were well managed, the floor areas needed for physician offices can be reduced. In a word, the core of hospital planning is optimizing the allocation and utilization of resources. Along with the ongoing healthcare reform, it's increasingly important to keep a tight rein on expenses.

There is a big concern on resource conservation. In China, medical insurance systems manage funds in a way backward. Take the payment for medical services as an example, it includes two parts: out-of-pocket personal pay and comprehensive pay by insurances. However, it is stipulated that "comprehensive pay is only applied to the expenses or fees charged during hospitalization". In order to get reimbursed by insurances, medical examinations necessary for hospital admission are often scheduled to be done after a patient is hospitalized. This inevitably leads to prolonged hospitalization and results in enormous waste of medical resources. Many common diseases can be cured in outpatient departments but the patients would like to be hospitalized due to the current medical insurance systems. Generally, the average hospital stay in China is 3-day longer than that in developed countries. Hence, China needs a reform on these policies and management systems. Given limited medical resources, we can't afford such a waste. Future development must be shortening unnecessary hospital stay, reducing inpatient beds, and increasing day-time beds for outpatients.

Lun Ge: West China Hospital is one of the largest hospitals nationwide. In recent years, there is a trend of building oversized hospitals. What is your thought regarding the scale of a hospital?

Yingkang Shi: *First of all, we shouldn't copy the hospitals built in other countries. We need to focus on the characteristics of our country.* China has her own characteristics. It would be inappropriate to say either "the larger the better" or "the smaller the better". The larger the hospital, the more challenges the administration faces. However, is it impossible to manage a large hospital? In my opinion, a large hospital doesn't always mean bad management. Of our hospital, the expansion is initiated by integrating the medical resources for miscellaneous disease and critical care. There is no intendance to compete with other hospitals on size. In order to utilize limited medical resources, a hospital focusing on miscellaneous diseases and critical care needs to be developed large in size to facilitate work efficiency. Another aspect is regarding the cost of administration. The smaller the hospital, the higher the percentage of administration costs in operation is. Therefore, we should conduct appropriate planning in the context of individual projects.

Lun Ge: In your discussion, what impresses me is that hospital decision-makers must be responsible to the society. We see your courage and philosophy. At last, could you talk about the most successful experience as well as the drawback seen in the development of your hospital?

Yingkang Shi: To conclude, there are four aspects upon our success. First of all, we have optimized the allocation and usage of limited resources.

Secondly, we make a practical and economically rational design for our hospital. There is no blind pursuit of a large size, foreign modes, or good lookings. The core of a hospital is its service processes responding to patient needs. Consequently, in a hospital design, nothing is more important than a reasonable service process.

Thirdly, architectural design and planning should take hospital administration and operation into consideration. It will help designers solve many contradictions and reduce costs.

Most importantly, we should use evidence-based ideologies in planning. For a long-term plan, it would be better to invite an interdisciplinary team to participate in hospital design and construction. Everyone can put forward suggestions from their specific professional perspectives. Our vision can then be widened.

Regarding the drawback, our planning is unable to keep up with all the changes. At the time of planning, we should consider things behind the surface and study the essential structure of operation and management in sample hospitals. This is of great significance. For instance, we should consider how they used their resources and what types of medical services they chose, etc. *As time goes by, there will be less resources worldwide and thus the global situations will be more severe: higher GDP, population aging, incurable chronic and geriatric diseases, and increased medical expenses. These are issues worldwide.* We should widen our vision and learn from each other regarding efficiency and resource saving.

Fig. 2.20 Yingkang Shi and Lun Ge

Lun Ge: You have sharp and distinct opinions on logistics and hospital administration. Thanks for your wonderful sharing (Fig. 2.20).

2.6 Open Mind, Update Concepts and Ideas, and Make Good Standards—An Exclusive Interview with Chen Wang

Time: March 2015—reiterated in May 2023
Location: Beijing Tiantan Hospital

Chen Wang: President of Tiantan Hospital affiliated with Capital Medical University; Vice President of Beijing Tongren Hospital affiliated with Capital Medical University; Vice President of the Beijing Hospital Association (Fig. 2.21).

Key sentences:

1. *From different perspectives, we both give priority to the importance of service processing in design decision making and agree that the convenience for patients and physicians is the core of design.*
2. *Of every factor, including technology and construction, the norms and codes should be kept updating. The standards of hospital construction must be adapted to the state-of-art of medical services and technology.*

Fig. 2.21 Chen Wang

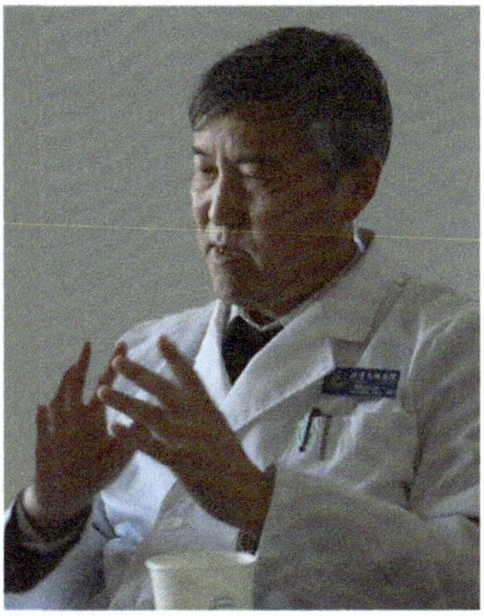

3. *Indoor air is more important than the outdoor air...It is very necessary to improve natural ventilation in hospital design. It is a key aspect that must be considered for medical buildings in the future.*

Lun Ge: Hello, President Wang! The new Beijing Tiantan Hospital have received a lot of attention. This project also experienced a lot of troubles. As the President, can you talk about your experience and feelings regarding it?

Chen Wang: I am a medical major. Architecture is a new disciplinary to me and my colleagues. So, we made great efforts on learning hospital planning and design (Fig. 2.22). As a client, we worked with experts inbuilding design and construction, to study design related knowledge such as planning, layout, and user flows. *Our methods were to develop design ideas based on the thoughts initiated by a selected design institute, then communicate medical knowledge with the designers, and finally realize the optimization of buildings and services through the incorporation on both sides.*

During the communication with architects, our opinions may be different from theirs. We recognized the importance of exchanging professional knowledge between architects and medical professionals. *From different perspectives, we both give priority to the importance of service processing in design decision making and agree that the convenience for patients and physicians is the core of design.* A lot of misunderstandings and contradictions have been resolved. *It would be ideal and much convenient for us if architects could better understand medical services, especially during the second-round of design focusing on service details.* If the details couldn't be fully represented, it would ruin the overall project. *From the perspective of healthcare providers, we concerned more about the specific design in detail.* This is the difference between medical architecture and other civil architectures, and the profound advancement that medical architects can contribute to.

Lun Ge: The problem does exist. Nowadays, there are many contradictions in medical design between hospitals and design institutes. The key problem is that the national

Fig. 2.22 Aerial view of Beijing Tiantan Hospital, Beijing

norms and codes of medical construction haven't been confirmed yet. Medical architecture in China is at a stage of transformation and we are all in the progress of exploring. As you mentioned earlier, in order to make architects better understand medical knowledge, some colleges of architecture are considering providing courses on basic medical knowledge. If so, it won't only reduce the conflicts in communication but also make it more convenient and efficient for both groups. So, President Wang, can you specifically talk about the key procedures in need of special attention during the design and construction of your hospital?

Chen Wang: I agree with you. The communication between us will certainly be smoother. Speaking of this hospital, the first key procedure was to set up and get a project approval. We did field trips to investigate hospital design across this country and globally. We reached a consensus on design after observing a lot of hospital layouts, at both macro and micro levels. For instance, we learned the concept of hospital streets and knowledge of appropriate story heights, signage systems and storage spaces etc. I hoped to have these concepts and factors applied to our hospital. *Meanwhile, we encouraged our staff to participate in mock-up research on operating procedures and logistics,* and publish articles on hospital's management of architectural design. We adopted some of the opinions and suggestions from our staff.

At the early stage of design, we thought about the relationship between the overall planning and future development, and how to handle the landscaping and external flows. We thought about humanized public spaces. For example, where to locate the restrooms? Should these be with corridors or on window side? Also, ICU expansion has become an inevitable trend. Then, how to balance the ratios of intensive care units to critical care wards and chronic disease wards? The most crucial point was that we must consider architectural design issues based on the characteristics of medical services and hospital development. Initially, many of us considered a high-rise building. However, I had concerns on vertical transportation. According to my previous experience, it would be a big pressure on vertical transportation in a high-rise inpatient complex. Since that, I suggested that the capacity of elevators should be well matched with the needs for vertical transportation. *We didn't make the building of Tiantan Hospital more than 10 floor tall and tried to make the interior transportation as horizontal as possible, which was the decision done at the early stage of design.* However, we still encountered troubles. For example, as a concept design, a symbol of metaphorical neural network was designed to be part of the building façade, to highlight the importance of neuroscience services in our hospital. However, applying this concept in reality was far from easy.

The second key procedure of design was across the entire project. It was the communication with varied parties. As mentioned earlier, we need to communicate with the architects first and then with construction agents. We focused on medical-related specialties, such as the design and construction of operation units, dialysis units, and MRI units etc. We should think ahead of all important specialties. I noticed the "Seven Systems" of a hospital building proposed by your team and couldn't agree more. I think these systems are more specific and systematic than prior ones. All medical specialties we consider for a hospital design can be found in these systems.

We also concern about the logistics management. Because of the large size and long-distance service trips in our hospital, we provide motor-trolleys for logistics use. We designed multiple stops to connect the PIVA Center and Nurse Stations, and optimized the network of logistics. Besides, we paid special attention to the construction of light–current systems and tele-medicine, in order to develop a digital hospital.

In addition, we focused on human-based measures. Public spaces such as cafes, lounges etc. help to develop healing environments. Amenities, such as easy access to inpatient units, self-service bathrooms, auto-flush toilets, and facilities for patient families, make the patients and their families feel being well cared. Last but not least, safety cannot be ignored. The entire hospital is fully covered by security cameras and automatic fire prevention systems. *Furthermore, building automation as well as energy saving and environment protection were all under our consideration.*

For the landscaping of our hospital, we tried to make a balance between attractiveness and functionality. Actually, it is becoming especially important to test the quality of air inside our hospital since the entire building is enveloped by glass curtain walls (Fig. 2.23). *More should be considered, including air quality standards, temperature and humidity requirements, and the standards used to select building materials.* Indoor air is more important than the outdoor air. However, attention paid to it is very limited. Many hospitals set up engine rooms by floor to support air conditioning. I think it has a big flaw. *Once we encountered with SARS and other airborne diseases, hospital-required infection rates across the floor could be extremely high. Hence, it*

Fig. 2.23 Main entrance of Beijing Tiantan Hospital, Beijing

is very necessary to improve natural ventilation in hospital design. It is a key aspect that must be considered for medical buildings in the future.

Lun Ge: You have done a lot of work and your idea is advanced regarding these practical problems. In fact, we also conducted some research in relevant fields. We would like to hear your thoughts on the SHAPE evaluation system for medical facilities.

Chen Wang: I am very glad to see that your team is working on this topic. *Due to the lack of specific national norms and codes in hospital construction, there are misunderstandings between hospitals and architects.* The norms and codes must be systematic and corresponding to medical specialties. We must intensively refine the coding systems to a certain extent. You and your team are experts in this field. We may conduct some research together with universities. Our hospital is still under construction. We would like to see if the SHAPE evaluation system could work as a warning system and guidelines for our hospital construction.

Last but not least, we must have a clear understanding of the trend of hospital development. The understanding should be kept updating. We update the concept of medical architecture through learning by doing. We need to conform to international trends but also be innovative. *We shouldn't set up too-much limit for architects but make them understand the law of hospital development. The nature of a hospital determines that the standards should be updated continuously. Of every factor, including technology and construction, the norms and codes should be kept updating. The standards of hospital construction must be adapted to the state-of-art of medical services and technology.*

Lun Ge: We are very willing to collaborate with hospitals to integrate theory with practice. Your sharing is great. We really appreciate (Fig. 2.24).

Fig. 2.24 Chen Wang and Lun Ge

2.7 The Soul of Hospital Administration—An Exclusive Interview of Lirong Zhuge

Time: January 2014—reiterated in May 2023
Location: The Preparatory Office of Shanghai Proton and Heavy Ion Center

Lirong Zhuge: Senior economist; former Executive Associate Director of Shanghai Proton and Heavy Ion Center Preparatory Office; former Associate Director of Shanghai ShenKang Hospital Development Center; former Chairman of the Hospital Architecture System Research Branch of Chinese Hospital Association; Director of Institute of Fudan Hospital support service & management (Fig. 2.25).

Key sentences:

1. *If a construction plan were issued without specific standards, it would have negative influences on project management, leading to excessive scales, standards, and investments (The Three Excesses). These would be an issue of government management.*
2. *We advocate natural lighting and ventilation and it should be considered on a case-by-case basis.*
3. *Many international companies provide beautiful design renderings to attract eye balls, but aren't address the needs of building functions well.*

Fig. 2.25 Lirong Zhuge

Lun Ge: Director ZHUGE, it is my honored meeting you. You are a pioneer in the field of hospital design and construction. You and your team have always been the leader of rationality and scientificity in this field. Recently, we collaborated with the Beijing Municipal Health Bureau on a research project focusing on the standards of hospital management and construction. I would much like to learn your experience and insights on this topic.

Lirong Zhuge: For 14 years, our work has been focused on the standardization of hospital design and construction. It includes four aspects: standardization, agent construction, quality, and honest construction.

Standardization: We conducted research on construction standards: the #18 Article of Fundamental Construction (2002), Project Performance Evaluation—Trial Version (2005), Primary Medical Construction Guidelines (2006), Construction Administrative Expenses and Control Standards (2006), Standards for New Construction of Tertiary General Hospitals in Suburbs (2009).
Agent construction: We conducted research on agent construction management to explore new mechanisms and modes of project management. *We focused on a group of key aspects including construction procedures, investment control, bidding management, financial management, contract management, schedule control, construction quality, and honest administration construction.* Besides, we compiled a compendium of agent construction project management.
Quality: We won competitions for major construction projects (2008–2013): 2 Luban Prizes for city-level hospital construction projects, 4 National Quality Silver Awards, 25 Shanghai Magnolia Awards for Construction Quality, 25 Outstanding Teams, 23 Outstanding Contributors, 14 Outstanding Organizers, and 85 Outstanding Builders.
Honest *administration* construction: We promoted ‘honest’ construction. We carried out the “Creating Double Excellence” activity with the municipal procuratorate to promote high-quality projects and excellent cadres, established a joint

meeting system, carried out anti-corruption education and supervised the major procedures such as project bidding, warning and prediction, to avoid corruption.

At the beginning, I was the vice president of Renji Hospital and in charge of hospital support services, construction, assets management, etc. After nine years work, I joined Shanghai Health Bureau in January 2001 as the Director of Planning and Construction. I was in charge of construction administration, focusing on medical institutions. Since then, I spent 14 years in the field of hospital construction and administration. I led the development of "*The 10th Five-Year Plan for Health Capital Construction in Shanghai*", which identified the functional orientation, development goals, and 28 proposed hospital projects in Shanghai from 2001 to 2005. After this plan got approved, we organized and formulated the general standards for hospital construction. *If a construction plan were issued without specific standards, it would have negative influences on project management, leading to excessive scales, standards, and investments (The Three Excesses). These would be an issue of government management.*

The '*Instructions for Strengthening Fundamental Construction of Municipal Hospitals in Shanghai' also referred as the '# 18 Article'* was developed in December 2002. The general standards for hospital construction were issued by the Ministry of Health of the People's Republic of China in 1996 and the standardized floor area per bed was 60 m^2. Many hospitals built on the standards were found to be overcrowded and in need of more lands and spaces. As a result, new construction projects have been suggested and intended to be built beyond the standards. Based on surveys and investigation, we established the Guidelines (the 18 Rules) and increased the standardized size of land and floor area per patient bed. However, we didn't set up the standardized budget for construction. The Guidelines were highly evaluated in practice. Since there were no standardized budget for construction, materials used in hospital projects varied greatly, say, some used stone, and some use paint. This leads the price difference between the two projects to be more than 1000 RMB per square meter.

In September 2005, Shanghai Municipal Government established a municipal institution for hospital development: Shanghai ShenKang Hospital Development Center (SHDC) (Fig. 2.26). Its focus was on government-owned healthcare assets and hospital projects invested by the government. I transferred from Shanghai Health Bureau to this Center and took the position as an Associated Director of SHDC. I was in charge of project investment and government-owned asset management. My team and I organized and edited the *Guidance for Strengthening Fundamental Construction of Municipal Hospitals in Shanghai. It was accepted and implemented in July 2006. There were five groups of standards, including the Construction Budget, Building constructions and sub systems, Equipment and materials, Energy efficiency, and Areas of structure.*

For example, for hospital buildings taller than 60 m and with two stories of basement, the budget was 5500 RMB per square meter. For general administrative buildings, the budget was 4500 RMB per square meter. Regarding hospital buildings shorter than 60 m and with one story of basement, the budget was 5400 RMB per

Fig. 2.26 Shanghai Shenkang Hospital Development Center, Shanghai

square meter. It also stipulates that the cost of weak electricity in hospitals is 250 RMB per square meter. The Guidance clearly stipulated that these standards would be fluctuant along with the change of material prices. In 2008, there was a continuous inflation of price in steel, cement, and labor. We conducted a study regarding the increase of construction costs and invited building designers, budget engineers, and hospital mangers to join us. Findings from this study led to an increase of construction budget, from 5500 to 6100 RMB per square meter. Once again in 2012, the market prices for construction materials increased a lot, and the budget was increased to 7100 RMB per square meter after a new study. In this way, an interaction mechanism was properly established between the budget and the market prices. The Guidance received approvals and supports from both the government and the Shanghai Municipal Development and Reform Commission. Similar adjustments have been made periodically in accordance with the market prices, to ensure the standards for construction quality remain the same.

In May 2001, the Shanghai Municipal Health Bureau and the Shanghai Municipal Health and Family Planning Commission established the Shanghai Health Infrastructure Management Center. It aimed at enhancing the efficiency and management of governmental investments in hospitals. It stipulated that all hospital projects invested by the government should adopt the mode of 'Agent Construction Systems', regardless of the amount of investment. At first, some hospitals doubted about it. After the implementation of 'the 10th Five-Year' Plan, they changed mind and accepted the mode of 'Agent Construction Systems.'

The main administrator of the 'Agent Construction Systems' is the Shanghai Health infrastructure Management Center. The Center edited the guidelines for managing "Agent Construction Systems" and jointly built a development office with each of the hospitals. Hospitals participate in the agent construction systems and work with the Center on specific project management tasks, including the clarification of construction needs, functional flows, service processes, facilities, equipment, investment and budget, etc. The three sections of management are as follows:

1. Designers should ensure that the investment required to build the proposed design is within the limits of budget approved by the government. It is clearly stated in the bidding documents for architectural design.
2. Design documents shouldn't be edited without a project experts' approval. For example, a proposed change on surface materials from paint to stone or ceramic tiles should be approved by project experts. Once a time, a hospital would like to change the design of its building facade into glass curtain walls and their proposal was denied by the experts after a project investigation.
3. There should be limits on equipment investments and material biddings. The bidding should be based on 90% of the estimated project costs. For instance, if the quotation is 5 million RMB, we bid at 4.5 million RMB. The rest of half million are saved for fixing possible misestimating. This method works out great. A construction project was initiated in 2007 and completed in 2010. Its gross floor area was 83,000 m^2 and the investment was 510 million RMB. Not only were the 'Three Excesses' avoided, but also 2 million RMB were saved. After the project was completed and its buildings were occupied, we selected a financial consulting company through bidding to conduct a comprehensive and systematic post-occupancy evaluation (also called performance evaluation). The evaluation was paid by a third party and the criteria of performance evaluation issued by the Center was applied. The project was evaluated in terms of function, process, convenience, security, user satisfaction, construction cost, financial management, scale and investment, etc. We used the evaluation results to learn lessons and share experiences.

Lun Ge: You and your team made so much effort. Clearly, it is a very solid and rational step in the field of hospital development in China.

Lirong Zhuge: We just completed the construction of Shanghai Proton and Heavy Ion Center (Fig. 2.27). It was a challenging project. There were a lot of difficulties in construction due to the requirements of building design. First, the structural subsidence was required to be less than 0.25 mm per meter per year. Second, of the machine room, walls were required to be 3.7-m thick, roof to be 2.7-m thick, and floor to be 2.9-m thick. The concrete crack should be less than 0.2 mm per meter per year for radiation protection. Third, the machine room for freezers and water pumps should be equipped with a floating floor to meet the requirements of microseism. Forth, the requirements for cooling water process systems were at a high level. The temperature of cooling water at the water inlet should be kept at 27 degrees and the pressure at 9.5 kg. Fifth, the temperature and humidity requirements for the machine room and system equipment were at a high level. Sixth, the requirements for power supply were at a high level. Seventh, we met a lot of difficulties in the development of pile foundations, concrete construction, and the installation of intricate utilities pipelines. Eighth, the schedule was very tight. In January 2009, we signed an international project contract. Based on it, the foreign party of this project would come to install and debug the equipment in three years. By then, we should confirm the accessibility of electricity, water, gas, and internet services. Successfully, ensuring project quality and safety, we completed the construction seven days in advance.

Fig. 2.27 Shanghai Proton and Heavy Iron Center, Shanghai

The experts from German spent seven days on the examination using laser devices, focusing on the 16,000 m^2 of built environment, including the machine room and the treatment room. They turned out to be satisfied on the outcomes and the project was delivered smoothly.

Looking back on this project, we made concerted efforts to build a completed team for high-quality organization and scientific management. We solved a lot of technical problems and successfully completed this project. I was the associate director of the project development office. During that time, I was the on-site coordinator and worked under pressure. This project received a lot of attention from the municipal leaders of Shanghai. If it weren't completed on time, we would be fined 0.02% of the total contract value for each of the days delayed, which would be 260,000 RMB per day. Successfully, we completed it seven days earlier. German experts approved its quality and accepted it with satisfaction.

Lun Ge: Could you please talk more about the objectives of SHDC?

Lirong Zhuge: Its role differs from that of Shanghai Health Bureau, which is in charge of hospital policies, planning, technology access, supervision on quality, and public health, etc. For example, a proposal of adding new patient beds or large medical equipment needs an approval from the Health Bureau. Whereas, SHDC was established during the reform of Shanghai Municipal Healthcare Institution to promote hospital development. The Center is responsible to managing nation-owned assets of municipal healthcare institutions as well as government-run hospitals. It ensures the implementation of government policies, guidelines, norms and plans concerning public health and hospital development, and directs the development of hospital design and construction. It is responsible to the policy decisions made by Shanghai Municipal Healthcare Institution, including asset management and contractor hiring. On the other hand, it is also in the charge of the investment, construction, operation, administration, and assessment of a hospital building project. The aim is to ensure a safe and effective operation of nation-owned assets and achieve the objective set for government-run hospitals. Meanwhile, SHDC promotes the reform of municipal public health institutions, improves the quality of hospital management systems and mechanisms, and helps to establish effective mechanisms of work incentive and constraint. It leads to more efficient use of nation-owned assets and improves the quality and efficiency of healthcare services. For 9 years, it has been actively exploring and practicing the normalization, elaborating a professional management of hospitals and hospital development, and guiding hospitals to follow the direction of public welfare development.

In 2013, the hospitals led by SHDC won 645 research projects funded by the Natural Science Foundation of China (NSFC) and collected 939 million RMB from the industry field for scientific research. In comparison, at the time (2006) when SHDC was established, these hospitals had only 114 research projects funded by the NSFC. To promote research, SHDC suggested these hospitals to pay more attentions on the growth of researchers and the connotation of healthcare services. We help to promote scientific research and improve the quality of healthcare services. In this regard, we formulated the standards to evaluate hospital presidents' work performance. The standards include 23 quantitative indexes and 3 qualitative indexes. These have been implemented for 8 years and highly evaluated by both the government and hospitals. Our projects on healthcare performance won the 'Policy Consulting Competition' hold by the City Government. We also won the 1st-level prizes of science and technology innovation in the competitions hold by the CHA (Chinese Hospital Association).

Lun Ge: You have gained so much success and experience. It's worth to spread the experience nationwide.

Lirong Zhuge: There is still a lot we can do. So far, the Public Hospital Administration of Shenzhen Municipality and the Center of Wuxi Hospital Administration Center have been established. The government of Shanghai and the Institutions of Public Services have been separated. SHDC is a public-service institution subordinate to the City Government. Experience sharing is led by the government. Our duty lies firstly in exploring the path of reform and concentrating on our tasks.

Lun Ge: Recently, we have been informed that although the Hospital Authority (HA) is nominally subordinated to the Health Bureau but at the same level of government management as the Bureau. This could raise problems upon the interactions of management.

Lirong Zhuge: The Health Bureau should focus on managing the whole industry of healthcare and directing healthcare practice, policy management, planning, and project approvals, etc. SHDC, on the other hand, is a public institution focusing on hospital development and administration. It undertakes government responsibilities to develop pragmatic public hospitals. Although nothing overlaps between the Bureau and the Center, they do share the goals of developing public hospitals and promoting the accessibility to healthcare resources in Shanghai. Our aims are to 'facilitate services, improve service quality, reduce the prices, and provide convenience'.

Lun Ge: I am really interested in the standards you have developed. Is it open to the public or confidential?

Lirong Zhuge: These documents have been openly issued in Shanghai. To sum up, we work on elaborative and professional management. The government of Shanghai adopted most of the opinions and suggestions from SHDC. For instance, the unit budget for hospital construction has been adjusted from 5500 RMB to 6100 RMB per square meter, and then 7100 RMB. We carefully and meticulously provide suggestions. The government trusts us. Both the Shanghai Municipal Development and Reform Commission and the Shanghai Municipal Audit Bureau believe that the administration of hospital construction is getting better. There are few phenomena of "Three Excesses".

To balance the healthcare resources in urban and rural areas, the government of Shanghai decided to build tertiary general hospitals in suburbs. It helps to improve suburban patients' accessibility to medical experts. SHDC was designated to develop the construction standards for the New Tertiary General Hospitals in Suburbs. It was completed in a relatively short period of time and approved by experts. Based on the Standards and with a unified project scale, planning, design and investment, all the tertiary hospitals were completed in three years from 2009 to 2012.

Regarding the hospital construction standards requested by the Beijing Municipal Health Bureau, I suggest that you and your team formulate specific rules and standards to clarify ambiguity. For example, what are the primary requirements for public hospitals? The size of a hospital building, its land use, and floor areas per bed should be clarified. These standards should be developed in the context of Beijing to address local needs. It asks for empirical research focusing on the specific needs of Beijing.

Shanghai adjusted the standards regarding hospital building sizes. Currently, the floor areas per patient bed in Shanghai are suggested to range in size from 100 to 120 m^2.

Lun Ge: Can you inform us the standards used in your inspection project? Are there any primary standards for quality control?

Lirong Zhuge: Actually, we ask all hospital projects larger than 10,000 m^2 should pursue the Shanghai Magnolia Awards for Construction Quality. We also encourage construction companies to compete for the Luban Prize.

Lun Ge: We consider 5 factors in post occupancy evaluation: safety, humanity, service process, physical environment, and efficiency. These factors are rated at 3 levels: qualified (required level), Standardized (average level), and Awarded (excellent level).

Lirong Zhuge: It's nice. The method of rating should be operation-friendly. You and your team have gone ahead of us in this respect. Attention should be paid on balancing the function of evaluation and the feasibility of evaluation. There should be an overall standard for project quality. Moreover, if a budget were exceeded, the excess should be controlled within 2% of the total cost. We prefer that each project can be completed within its budget. Limits must be set up for extras. Standards are developed to imposing restrictions and thus shouldn't be too broad otherwise things may be out of control.

Lun Ge: Of a hospital project, the review process in Beijing initiates at the level of project proposal and continues on the report regarding project feasibility and biddings for architectural design and design development. In order to emphasize on the importance of Beijing Municipal Health Bureau in this process and highlight the importance of pre-design research in healthcare planning, we plan to add work on the evaluation of project programming. We suggest inter-level research on project feasibility between the level of schematic design and the level of design development.

Lirong Zhuge: In Shanghai, hospital projects need to go through a level of pre-assessment before submitting project proposals. At this level, the information on project feasibility is asked, including project planning, construction, structure, environmental protection, and budget. Project reviews are not conducted by the Health Bureau of Shanghai but SHDC. Based on the reviews done by SHDC, project proposals are then submitted to the Shanghai Municipal Development and Reform Commission and the Health Bureau for pre-evaluation.

Lun Ge: According to the general procedures of hospital construction in Beijing, the Development and Reform Commission and the Construction and Planning Bureau are authorized respectively to review project proposals and issue approvals for land use, including the percentages of floor area utilization and green areas, etc. No institutions focusing on healthcare facilities are authorized until the Institution of Beijing Hospital Management is included in the procedures.

Lirong Zhuge: In Shanghai, we have the Institution of Hospital Management and the Institution of Construction and they are individually independent. After the reform, SHDC is in charge of the review on hospital project proposals.

Lun Ge: The absence or deficiencies of preliminary work leads to resource waste. The phenomenon of "Three Excess" needs a lot of attention. One of the purposes of our current study is to strengthen the role of Health Bureau in this regard.

Lirong Zhuge: I suggest a focus on elaborative and professional administration. To ensure the quality of a proposal review, a professional and elaborative administration should be developed to eliminate the "Three Excess": excessive investments, excessive standards, and excessive scales.

Lun Ge: Our study also aims at facilitating elaborative and scientific hospital management. Building construction should follow the standards. However, it is hard to define the standards. Are there any empirical standards to follow?

Lirong Zhuge: When you and your team developed the standards, you have the standards in place for managing government's investment on public health. In the future, project review and building construction wouldn't be separated. There is no need to adjust the proposal of overall investment upon the completion of construction.

Lun Ge: What is your opinion regarding the plan of adding assessments on the technical process of medical services and equipment systems before reviewing the document of design development?

Lirong Zhuge: After approving a report on project feasibility, additional reviews on its design of functional flows and equipment may lead to intensive adjustments. It could be quite troublesome if the scale and investment of a project changed. The government has clarified that theoretically: we are not allowed to adjust the scale or investment of a project at the level of design development once the report on project feasibility has been approved. Based on these, it is recommended to add assessments on functional flows and equipment systems before approving the report on project feasibility.

Lun Ge: Nowadays, many hospitals claim to be 'a top-level hospital worldwide.' What do you think about this kind of slogans?

Lirong Zhuge: What is the definition of 'a top-level hospital worldwide'? In my opinion, we might claim that our medical technology is at 'the top level worldwide.' Whereas, it's inappropriate to say that in regard to our hospital buildings. At present, hospital buildings in China are built to address the needs for primary healthcare, with an emphasis on hospital functions, humanity design, and appropriate and safe environments. This phenomenon is differing from that in foreign countries since we have a large population.

Lun Ge: This is also because of the absence of strict standards for project reviews and definitions. Many may take it for granted.

Lirong Zhuge: That's definitely not allowed. For instance, we should clearly state our opposition to building large glass curtain walls or metal frames in hospitals. I think it's still necessary for hospitals to focus on public needs for primary healthcare in terms of the environment, safety, convenience, anti-infection, etc. Moreover, is the claim of 'a top-level hospital worldwide' made in comparison with America or Australia? Our economic conditions are different from theirs. Hence, the expression of 'a top-level hospital worldwide' is intrinsically flawed and shouldn't be used. Nevertheless, it's fine for a for-profit hospital to claim to be equipped with medical devices which are at the top level worldwide. But there is no need for public hospitals to advocate themselves to be at the top level worldwide.

Lun Ge: Do you think there should be some specific standards focusing on the natural ventilation and lighting systems in a hospital? Many hospitals have dark and enclosed rooms with no access to natural ventilation and lighting. What's worse, there are no specific design requirements in this regard.

Lirong Zhuge: *We advocate natural lighting and ventilation and it should be considered on a case-by-case basis.* It is very hard for designers to address the needs for natural ventilation and lighting in outpatient departments. Nonetheless, many hospitals build an atrium in their outpatient departments to address the needs. We suggest using atriums and patios in design to reduce the number of dark rooms.

Lun Ge: I studied in Europe. All major hospital rooms designed by my instructors had good natural ventilation and lighting. These hospitals were small in scale. Compared with them, hospitals in China are much bigger and thus providing natural lighting and ventilation to all the rooms is quite difficult.

Lirong Zhuge: Take our hospital as an example, the number of outpatients per year is counted by million. In contrast, a general hospital in Australia has about one thousand beds and accommodates 340k outpatients annually. Things are very different. For instance, in Shanghai, there are 8 hospitals visited by more than 10k outpatients daily.

Lun Ge: In China, there are many design companies from foreign countries. They collaborate with local designers and hospitals on hospital projects, such as the project of Zhongshan Hospital. What's your opinion on that? What do you think about these design companies?

Lirong Zhuge: We welcome international design companies to China. In hospital biddings, it should be clearly pointed out that hospital buildings may not have glass curtain walls in design. *Many international companies provide beautiful design renderings to attract eye balls, but aren't address the needs of building functions well.* Many domestic companies are very good at designing functional layouts, but their building facades are relatively simple and rigid. However, some have done very well in recent years.

Lun Ge: Which hospital in China do you appreciate and think as a success?

Lirong Zhuge: A public hospital provides primary healthcare services. Responding to patient needs, it should provide good ecological environments with natural ventilation and lighting. For a hospital, it's unnecessary to claim to be at the top level worldwide. Likewise, it's not the case that only famous international design companies can design good hospitals. Up to now, hospitals that do well in this aspect, as I know, include the Shanghai Public Health Clinical Center, the Children's Hospital of Fudan University, and four new tertiary hospitals in the suburbs of Shanghai. Overall, these hospitals are well-designed with centralized layouts. We invited many designers and received 14 design proposals. Those from Shanghai Institute of Architectural Design and Research won.

Lun Ge: Thanks very much for your time. I have benefited a lot from this interview.

Lirong Zhuge: It's just my personal experience. Please kindly comment and suggest if there is anything inappropriate. Thank you (Fig. 2.28).

Fig. 2.28 Lirong Zhuge and Lun Ge

2.8 Building an Integrated System for High-Quality Hospitals—An Exclusive Interview with Jianzhong Zhang

Time: January 2015—reiterated in May 2023
Location: The meeting room of Shenkang Corporation, Shanghai

Jianzhong Zhang: Senior Economist, National Certified Supervisory Engineer, Royal Chartered Surveyor, US Certified Engineering Cost Engineer, Director of Shanghai Hospital Construction Management Center, CEO of Shanghai Shenkang Hospital Development Center (SHDC) (Fig. 2.29).

Key sentences:

1. *We shall "indigenize" experience from foreign countries and conduct research and exploration in the context of China.*
2. *Promoting the management mechanisms enhances the quality of operation. Operations conducted on a "large platform" facilitate monitoring and respond to future inquiries.*
3. *It is necessary to integrate and assemble the factors of pre-design planning and post occupancy evaluation ...We need to formulate a good mechanism of management. It's necessary to gain a thorough understanding of a hospital's organizational background and its construction systems.*

Fig. 2.29 Jianzhong Zhang

Lun Ge: Nice to meet you. Shanghai Shenkang Hospital Development Center (SHDC) is the national pioneer in hospital construction. SHDC has developed a relatively matured mode of construction management through years of practice and accumulated a wealth of experience. You have been working in this field for many years and developed a lot of distinctive insights and practical methods. Would you please share with us your thoughts and experience in this field?

Jianzhong Zhang: Before the 10th five-year plan, the construction of hospital buildings in Shanghai were conducted by individual hospitals. A hospital building is very different from other types of buildings because of its medical properties, which differentiate medical and general architecture. *To build a hospital, it is required to integrate the two types of architecture and thus far from easy. Furthermore, every hospital has its unique functions, positioning, and requirements. These differences then determine the distinct ways of its planning and construction. Technologies grow very fast. However, we still lag behind in cognizing and implementing a framework for hospital project positioning and systematic development.* For instance, in the era of big data, everyone talks about building "green hospitals". Nevertheless, we couldn't set up a full set of standards at one time, for green hospital construction, building management, and medical services. We may explore it step by step.

First of all, in the life cycle of a hospital, its functional positioning and daily operations are interrelated. If these were separated, it would result in continuous building renovations. *At present, hospital construction in China is still in need of a clear and thorough understanding of hospitals, in terms of project positioning and properties.* Inadequate investigation results in deficient cognition and understanding. As a result, hospital buildings with incompatible characteristics would be built. Inappropriate design of medical-service processes may cause a big waste of labor, materials, and financial resources.

Especially, our research on hospital safety is very limited. We don't have a thoughtful plan interrelating the safety issues in terms of environment, equipment, and medical instruments. There is also an issue of building safety. *During the life cycle of a hospital building, we should take the considerations of building safety into account.* If we didn't consider these issues as a whole, the longevity and efficiency of a hospital building would be negatively influenced. We need to classify research and exploration in order to figure out the whole system of hospital design and construction. *We shall "indigenize" experience from foreign countries and conduct research and exploration in the context of China.*

Secondly, it is regarding the management mechanism of SHDC. *Upon the establishment of* SHDC, *we defined its aims. It should be professional, normative, elaborating, scientific, and informative.* At that time, we sorted out the factors associated with the process of hospital design and construction. Starting from the phase of project positioning, we participate in the project management and conduct work summaries and reviews upon the completion of each phase. We engage in the entire process of project development.

It's quite necessary to sort out and summarize work by phase. This type of systematic work promotes the proficiency of management. It includes the research on general project management and the administration of agent construction system. We edited the *Outlines and Guide for Construction Project Management in Shanghai* in order to address the gaps among professions. It is of pioneering significance in that a framework has been developed for the entire industry of hospital construction management.

The third aspect is focused on "Double Strictness and Double Excellence". "Double Strictness" refers to the restrictions on both project scale and investment. "Double Excellence" means to combat corruptions and demonstrate an honesty management. We conducted research to analyze the risks of project development and identify the factors in need of a special control. Based on this research, we developed the software "Shenkang Project Management, Risk Prevention and Control". Firstly, it provides a platform of management and serves as a bridge between project participants for information sharing and communication on the key factors of each project phase. Secondly, it presents a completed process of project development. The process is divided into 6 stages, 28 sections, and 200 steps. We set alarms at each of the stages, sections, and steps in order to keep a strict control over project operation and management.

It reminds us what to do at each stage of a hospital project. We are in need of professionals who know the entire process in an all-round way of hospital development. By using this software as a platform, we help project managers to understand the specifications and operation requirements on hospital construction in an all-round way. This promotes staff training and education.

Fourthly, it helps to conduct a timely and rigorous control over project investment. On this platform, an application should be submitted in advance before any changes could be made. That includes the reason for change and whether the change is rational, etc. Therefore, we can guarantee a more normative management of projects and a better control of the investment and the project scale.

This platform gives proper advices to different departments and helps to formulate the standards of management. By operating and managing this platform, all the managerial personnels involved in one project can coordinate and cooperate with

Fig. 2.30 Inpatient building of Ruijin Hospital Affiliated with Shanghai Jiaotong University, Shanghai

each other. This ensures smooth management and high-quality services. This platform facilitates the implementation of "Double Strictness and Double Excellence" (Fig. 2.30).

In addition, we conducted research and improved the standards of construction. It includes a research project sorting out the standards of electrical systems and informatics construction. In addition, we have formulated the basic configurations and criteria of project development, including a simulation of project functions. For instance, what exactly would the volume of outpatients be? It's necessary to formulate a yardstick based on functional simulations. Meanwhile, we should be aware that the internet influences both the process of medical services and the implementation of building construction.

As we mentioned earlier, there are "three restrictions" on construction projects. These undoubtedly set up high-level requirements for project management. As an agent construction company, SHDC emphasizes the importance of project management. *It aims at controlling the projects in terms of quality, progress, investment, civilized construction, safe production, and individual behavior.* These procedures not only ensure the quality of project, but also improve the maneuverability of project. Management issues involving multiple parties can then be reduced.

The fifth aspect involves the management of project operation. *Promoting the management mechanisms enhances the quality of operation. Operations conducted on a "large platform" facilitate monitoring and respond to future inquiries.* Advent of the big-data era has brought us a variety of opportunities, challenges, and new

Fig. 2.31 Songjiang District Hospital of Shanghai First People's Hospital, Shanghai

thoughts. *If aforementioned methods of management can be supported by the big data, it will be very beneficial for the entire project. It is what we need to consider for the future* (Fig. 2.31).

Lun Ge: You have very rich experience managing hospitals. What do you think about the preliminary planning and post occupancy evaluation regarding hospital projects?

Jianzhong Zhang: *The prior construction to a great extent lacks pre-design planning and post occupancy evaluation.* Hence, I have come up with three advices. The first one is regarding construction management. Individual departments of the central government are in charge of different aspects of project development. So, we have to integrate them as a whole. For instance, the Committee of Development and Reform is in charge of project initiation whereas the Ministry of Construction is in charge of project construction. *It is necessary to integrate and assemble the factors of pre-design planning and post occupancy evaluation.*

The second advice goes to project construction. We need to obtain a full understanding of individual groups involved in the construction. We also need to specify a hospital's medical functions and the entire process of medical services. Only by figuring out the functions and the service process could we understand the interrelation of project needs. Then we could make better project planning. On the other hand, in terms of evaluation, we should develop an evaluation system for the entire project and individual phases. Only in this way can we make a comprehensive project evaluation both as a whole and in each phase.

Thirdly, the utilization of hospital equipment and devices are always passive. Most of them don't have the required quality. In the era of internet, we should change this passive situation and gain the power to control it. For example, we should figure out clients' requirements by group discussion online or on a certain platform and then find ways to implement the required equipment and devices. Designers are in need of a platform for interaction. The requirements for project modification are often incomplete or un-thorough. These eventually lead to low-quality projects. *The quality we discuss here is not of one product but regarding the entire system.* As a consequence, it's essentially necessary to promote our ability to control it.

Here are our plans. Firstly, *we need to formulate a good mechanism of management. It's necessary to gain a thorough understanding of a hospital's organizational background and its construction systems.* Secondly, we need to put particular emphasis on the development of green buildings and the utilization of information technology, etc. More close attention is needed to analyze the potential problems and their influences on hospital development. *Thirdly, we should research both hospital functions and operation procedures and then refine relevant regulations and standards.* Whether or not a hospital pays attention on its unique characteristics and the state-of-art in medical technology influences the direction of development. Fourthly, it's necessary to indigenize international experiences in a local context to integrate varied types of management.

In terms of pre-design planning and post occupancy evaluation, regarding project functions, we should firstly have an internal review and then offer enough space for hospital users to conduct in-depth discussion. Afterwards, we would once again sort out hospital functions and make necessary adjustments in accordance with the mandatory norms set by the government. *Upon completing the refinement, after further discussions with hospital users, we would develop a list of project tasks.* The list of tasks is particularly initiated by us and then open for bidding. After bidding comes experts' review. The design institute then integrates the comments from reviewers and makes necessary adjustments. Subsequently, we submit a proposed design program for approval. Upon its approval, we will move on to the stage of design development. Once the hospital's functional positioning is set up, we move forward to develop construction documents with selected design institutions.

Furthermore, we have established a platform for group discussion. This platform invites every participant in the project and facilitate communication. Questions would be identified through discussion. We will sort the questions, find solutions, and then give feedbacks on the platform. *A shared platform greatly promotes and facilitates project management.*

Lun Ge: After the interviews with you and Director ZHUGE, I gained an in-depth understanding of SHDC's management system. You have developed such an advanced system in practice. Once again, I would like to suggest everyone in the field of hospital design and construction to learn from you. Thanks again for sharing (Fig. 2.32).

Fig. 2.32 Jianzhong Zhang and Lun Ge

Chapter 3
Interviews with the Vice Presidents Focusing on Construction Management in Hospitals

3.1 Hospital Construction is a Range of Controls from Fundamental Construction to Logistics Management—An Exclusive Interview with Mei Chen

Time: January 2015—reiterated in May 2023
Location: The office of Shanghai Sixth People's Hospital, Shanghai

Mei Chen: Vice president of Shanghai Sixth People's Hospital, Senior engineer, Deputy Secretary General of China Hospital Association's Building System Research Branch, Vice president of the Logistics Management Research Institute of Fudan Hospital (Fig. 3.1).

Key sentences:

1. *It requires a professional background and scientific methods to address user needs and meet the requirements for project continuity, security, energy saving, consumption reduction, and less or no equipment breakdowns.*
2. *Preliminary planning and POE are crucial to hospital construction. In the overall management, we should hold a holistic view on various factors associated with the preliminary planning, project verification, and construction documents.*
3. *When a hospital is occupied, the quality of equipment, such as valves of pumps, catches attention. Once there is any equipment breakdown, its influence can be fatal to the hospital. Therefore, during the construction, we should pay attention to the equipment, make sure it meets the standards for energy saving and safety. Construction and practical operation should be considered simultaneously.*

Lun Ge: Nice to meet you, President Chenmei. As a female president in charge of hospital construction and logistics management, you have given a full play of feminine specialties and a pursuit of perfection. You obtained fruitful achievements

L. Ge and Z. Wang, *Thoughts on Hospital Design and Construction in China*,
https://doi.org/10.1007/978-981-97-0818-5_3

Fig. 3.1 Mei Chen

in the construction management of Shanghai Sixth People's Hospital and accumulated abundant experience. Could you share with us your thoughts on hospital construction?

Chenmei: I have been engaged in construction and logistics for over two decades. I majored in industrial and civil engineering. After participating in several projects of hospital construction and relocation, I was impressed that hospital buildings were far more distinct than other types of buildings.

First of all, buildings need to meet user needs. Hence, communication with the users is particularly important. I value the communication with physicians and staff in the frontier, through which I get to know the deficiencies of a construction design and the process of medical services. We make improvements by conducting continuous refinements in practical operation. What's more, similar defects could be avoided in future projects. Fortunately, during these years working in this hospital, seeing the rapid development of medical technology as well as the constant building reconstruction and extension, I came to know what is rational for medical services and gained a deep understanding in this regard. Thanks to the big projects I participated in and the flexural processes of communication with doctors and nurses in the frontier.

During the continuous growth of a hospital, buildings have to be updated with time to meet patient needs and the new requirements for pipelines and medical equipment (Fig. 3.2). During a hospital's relocation over two decades ago, we obtained an assistance from the Japanese Jeka program and received a set of medical equipment worthy of about one hundred million yuan. However, at that time, many of our buildings were completed so we had to renovate them in order to install the equipment. We tried our best to ensure that the equipment can fit in with as less modifications and costs as possible. The experience accumulated during this project was so helpful that we would be more confident in managing the subsequent installation of medical equipment.

Fig. 3.2 Shanghai Sixth People's Hospital, Shanghai

Secondly, Shanghai Shenkang Hospital Development Center (SHDC) requires "Double Strictness and Double Excellence" for hospital construction management. As far as I'm concerned, it's applicable to keep the scale of a hospital under control since governmental examinations are so stringent that the project would be rejected upon any disqualification. However, it is difficult to follow the specific construction standards during the long process of project control. From the initial establishment of a project to its completion, it takes at least three or up to five years. During such a long period of time, balancing and controlling the prices of labor, materials, machinery, and the discrepancy between the investment valuation and practical needs is truly hard. *It requires a professional background and scientific methods to address user needs and meet the requirements for project continuity, security, energy saving, consumption reduction, and less or no equipment breakdowns.* Via constant communication and adjustments, the aim is to achieve a good performance and keep everything in budget.

Lun Ge: You may have a certain understanding of our scientific research on preliminary planning and post occupancy evaluation (POE). Would you please offer some suggestions?

Chenmei: *Preliminary planning and POE are crucial to hospital construction. In the overall management, we should hold a holistic view on various factors associated with the preliminary planning, project verification, and construction documents.* The specific items include air-conditioning, service procedures, and construction costs. Furthermore, we need to consider how to deal with project delivery upon construction completion, such as the run-in period after occupancy and budget evaluation. These are essential in the routine of construction management. *Integrated construction management runs through the entire process, from preliminary planning to intermediate stages and then POE. Only when these were well done, would the work be perfect.*

We have initiated post occupancy evaluations. In Shanghai, POE is required for hospital projects invested by the government. SHDC has stringent controls on the evaluations. On one hand, SHDC keeps an eye on preliminary formalities to ensure that the projects conform to relevant national stipulations. Then, preliminary budgets were estimated by the Center to ensure that the costs don't go beyond the budgets. After that, any additional investment would be reviewed to see if it were necessary and followed the relevant formalities. On the other hand, patient and staff's opinions are taken into consideration while overall evaluations are made regarding the process of medical services, hospital functions, technologies, comfort levels, and medical indicator changes. As a result, meticulous works have been done in this regard, reflecting overall considerations of the construction.

Here is a small tip. During the construction of a hospital, I paid special attention to the quality of logistics equipment (Fig. 3.3). The selections on central air conditioning systems, heating/ cooling modes, medical gas systems, and water pump valves should follow certain standards. *Any failures of the logistics system could be fatal for the hospital. Therefore, in regard to hospital construction, we should make sure that all the equipment is safe for operation, meets the requirements for energy efficiency, and can run well to ensure the continuity of medical services.*

Lun Ge: You participated in three projects of new hospital construction and several projects of hospital reconstruction and extension. What are the key points of project success?

Chenmei: First of all, building construction should follow the overall objectives of hospital development. That's, hospital positioning guides the planning of construction. *It's critical to think over how to develop the hospital and choose an appropriate mode of operation.* For example, if a hospital were proposed to be a research hospital, the relevant factors of scientific research should be considered in planning.

Secondly, preliminary planning is of great importance. It's necessary to specify the scale of a hospital and consider the largest volume of emergency patients, outpatients,

Fig. 3.3 Courtyard of Shanghai Sixth People's Hospital, Shanghai

and inpatients. Objectives on "medical services, teaching, research, and prevention", etc. should be comprehensively considered in the preliminary planning of a hospital.

Furthermore, full attention should be paid to schematic design, which decides the overall mode of hospital operation. The proportion of floor areas should be decided by department and based on actual needs. If there were clinical centers or departments, we should consider what kind of devices should be provided, how to install them, and what the procedures should be, etc. Only by figuring out all of these, could we determine the basic layout of a hospital building and thus know how to distribute the logistics equipment. We should be fully prepared regarding the schematic design for the fear of rework.

In terms of administration, the coordination of all parties helps to reduce modifications and rework. *We should encourage the directors of clinical departments to participate in hospital construction so that they could gain a more perceptual understanding of construction.* By doing this, we can greatly improve the efficiency of construction and create favorable environments to carry out the "Double Strictness and Double Excellence". Moreover, the process of design is critical. Decades before, the basic mode of medical services was 'patients follow doctors' and thus patients rushed everywhere in the hospital for medical services, examinations, treatments, and hospitalization, etc. This mode is fine for small hospitals since it takes just a few minutes to go around the entire hospital. Nevertheless, it is easy for patients to get lost and exhausted in the wayfinding in a large hospital. *Future hospitals should be patient-centered.* A hospital may focus on selected disciplines and assemble the services of emergency care, outpatient care, examination, and inpatient care. Home-like atmosphere should be provided to promote care. In other words, it's going to be a medical center with an emphasis on selected disciplines.

Finally, I would like to talk about the professional background of construction. I think the biggest advantage that I have in managing hospital construction lies in my professional background and genuine attitude in communication. Whoever in charge of the construction should be of a professional background, or at least there are some construction professionals in the team. Team members can then understand both building construction and medical services. In this way, *the building can better meet user needs.*

Lun Ge: Were there any design or construction regrets found upon building occupancy?

Chenmei: In fact, a lot of problems occurred after the buildings were put into use. Let us take the design of pipelines for example. There are various pipelines in a hospital, due to the intricate procedures of medical care and the functions of medical buildings. Particularly, the route and materials of sewage conduits require special attention. The sewage treatment station in our hospital is located at the lower underground level and the sewage is drained into the city sewage network by water pumps. Since the underground water level is higher than the outfall level of our sewage treatment station, waste water would flow backward once the city sewage network is overloaded. Therefore, it's crucial to add a check valve at the outfall. Furthermore, we should calculate the length of vertical outlet pipes to decide suitable pipe materials accordingly. Once there was a sewage pipeline burst because of inappropriate pipe design. After that, we successfully avoided all pipeline burst by replacing the pipeline elbows with those made of tougher materials and with better impact resistance. These are the hidden problems in construction engineering. We should be as careful as possible during the process of design so that a lot of extra work could be avoided.

Lun Ge: Could you talk more about the design of your project brief and the decision making regarding medical-service processes at the stage of preliminary planning?

Mei Chen: *We designed the project brief by ourselves. This is of great importance and associated with the architects' work.* Their design was based on the project brief. It's the hospital that leads the project and the brief reflects hospital needs. Based on our project brief, the designer team further clarified the needs of medical procedures and processes. We conducted constant communication with the designers and updated the brief correspondingly. Therefore, the coordination and communication of all the parties are particularly important. Design drawings should be as detailed as possible, including the relevant indicators. Only if these are appropriately performed, will the overall processes of construction go smoothly. Regarding architectural decoration, we prefer a simple and elegant modes, which should be convenient for cleaning and disinfection. The majority of our emphasis is on the aspects of medical services, equipment, and quality.

Lun Ge: What should we do to guarantee the quality of a hospital building?

Mei Chen: A building is an integrated product. Therefore, first of all, it's necessary for hospital staff to participate in hospital construction management. The "Agent Construction Systems" are applied in Shanghai. Secondly, *project examination and the delivery upon construction completion is very important.* The 'user' departments should participate in the examination. Initially, we mentioned that a building should meet user needs. Therefore, the departments, the actually users, are the best qualified to make comments. They offer opinions based on their daily use of the building, including things as trivial as if each medical-gas plug can be easily used, whether the gas supply is normal and the air quality is up to standards. So do the comments on water temperatures and air conditioning, etc. These are the indicators of POE as well.

The government office of construction management, the supervision party, and the 'user' departments have their specific standards and priorities regarding the project examination and acceptance. Only when all the parties are satisfied, can we assure that the building is up to our standards. Generally, we repeat the examination process for several times and make constant rectifications until the building meets all the requirements.

Lun Ge: Thanks very much for your comprehensive and insightful sharing! (Fig. 3.4).

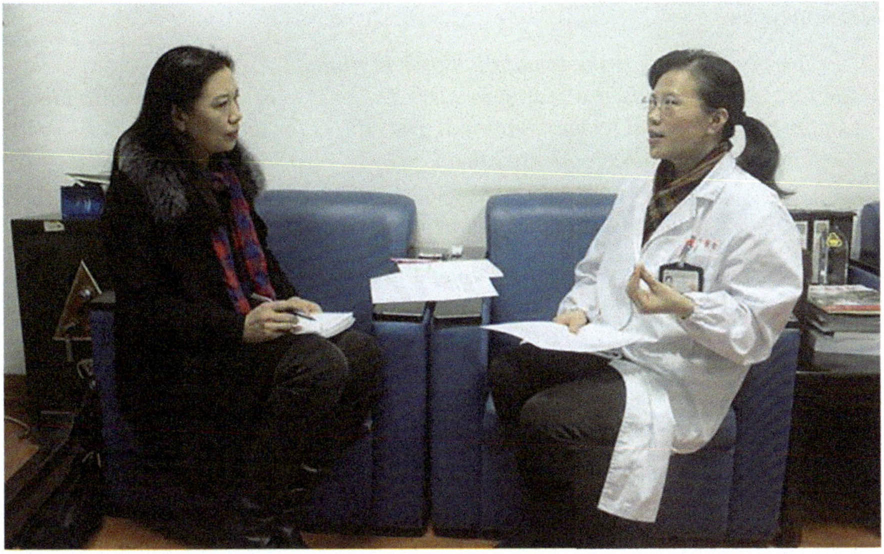

Fig. 3.4 Mei Chen and Lun Ge

3.2 Hospital Design Starts from a Comprehensive "Diagnosis"—An Exclusive Interview with Jianhua Dong

Time: January 2015—reiterated in May 2023
Location: A coffee house, Beijing

Jianhua Dong: Former Director of the Construction Department of Beijing University's First Hospital, Vice President of Beijing Hospital Construction Association (Fig. 3.5).

Fig. 3.5 Jianhua Dong

Key sentences:

1. *Project positioning is the foundation of everything.*
2. *In fact, evidence-based research in early design is necessary in order to identify appropriate medical technologies and procedures for the design. Otherwise, it's essentially impossible to identify them.*
3. *It's necessary for a hospital to make a thorough and indigenized analysis of itself, leading to a clear understanding of its future development.*

Lun Ge: You are a walking encyclopedia in the field of hospital design and construction. I always look up to you. You are proficient in various disciplines. You also have a systematic ideology that always guides you to find scientific solutions. You can analyze and solve problems at macro, medium, and micro levels. I think you are comprehensive, profound, and insightful. I am extremely curious how you have become what you are now. Could you share your experience?

Jianhua Dong: In 1983, I was in charge of my first project, a Maternal and Child Health Care Hospital. After that, I participated in a variety of hospital construction projects, including an MRI center, buildings for teaching and research, a surgery center, a medical center, and an outpatient department. Asked why I have the privilege to work with so many hospitals as a consultant, I would say it is not necessarily associated with my developed social network, but the other ways around. That's related to my experience regarding how to consulate to address the needs of hospitals. *What our hospitals need the most is preliminary planning.*

It's neither the case that no one notices the importance of preliminary planning in China, nor that no one emphasizes on its importance. As a matter of fact, in the latest version of the book '*Construction Standards for General Hospitals*', there's even a specific Chapter devoting to medical technologies. Why would we do that? *We have been studying preliminary planning for a long time, which includes two dimensions: practical and theoretical.* What I have engaged in is the former. Professor Lun Ge have done enormous theoretical research and applied the findings in practical projects such as Taida Hospital. At most of the time, I participated in hospital projects during the phase of preliminary planning. *I have learned that many hospital projects are in urgent needs of a thorough and indigenized preliminary planning.* This is what has impressed me the most.

Speaking of "indigenization", I mean that we must take into consideration is the actual situations of a hospital. Neither shall we aim too high, nor bury our head in the sand. Only by making a thorough diagnosis of its practical situations, could we know what to do the next. A hospital is a system and should always be the core of its construction planning, no matter who is the planner, say, a design institute or a consulting company. *The hospital itself should be aware of the large framework of medical services it offers.* Then, I will help to develop the framework of its preliminary planning.

Focusing on the development of project briefs, I participated in many hospital projects and wrote the first hospital design brief in China. The brief was for our surgery center and highly evaluated by Director Li from the Ministry of Health.

Since then, all hospital construction projects are required by the Ministry of Health to submit a project brief on design specifications and contents.

In a project brief, the purposes, profiles, plans, conditions, and requirements must be clarified. In other words, a project brief is not ready-made, as far as I'm concerned. No matter how a hospital would be expected, the design should always be based on its practical situations. However, nowadays over 80% of the hospital design institutes in China develop project briefs by transposing an existing one and run with rough knowledge on hospitals. After grabbing some requirements of a hospital or receiving requests from a client hospital, a hospital manger or designer may then draft a project brief. Such brief development generally runs in three rounds and 80% of the briefs are completed in this way.

Therefore, I especially hope that hospitals can set preliminary planning as an independent section in project development. What's more, it should be initiated before the project received its approval. *Preliminary planning should be required and given special attention before construction.* No matter the planning is led by the hospital, hired experts, or a professional company, it is necessary and should be done in advance. It would be too late if conducting preliminary planning during the period of brief development.

As you can see, at the international level, evidence-based design and research has been set up as an independent stage of preliminary planning. First of all, evidence-based design has specific design goals, sources of evidence for design, and critical interpretations of the evidence data. Based on these, designers then move on to the phase of design innovation and creation. They propose design hypotheses, implement and simulate design ideas, and compare the results, which correspond to the phase of schematic design. Lastly, an industrial project review would be conducted. These steps can make a closed cycle.

The procedures that a hospital construction project in China needs to go through are similar. First of all, in order to issue a project approval, a general project review is required and aimed at problem solving. Facing a lot of project contents, experts are asked to provide systematic solutions in a short period of time, without evidence-based research. How scientific do you think the solutions would be? There are two reviews. The first one is Proposal Review and the second is Feasibility Review, which requires a schematic design. In my opinion, these reviews have been handled in unscientific ways. *I hope a project approval can be based on evidence verifications instead of expert reviews.*

According to my experience, we should first understand the overall situations of a hospital's administration, which includes how it has been run and the direction of its future development. Every hospital is different and cannot be replicated. Therefore, only by gaining a full understanding of it, can we have a sense of innovation and be able to integrate advanced ideas into its development, leading to a successful project. *Therefore, project positioning is the foundation of everything.* Only based on that, could we formulate a project brief or a report of feasibility study.

I think it's wise for a hospital to get things done with social supports. Only in this way can we make up for shortcomings. In other words, we have devoted a special Chapter in the new book 'Design Standards for General Hospitals' to discuss medical technologies and procedures. *In fact, evidence-based research in early design is necessary in order to identify appropriate medical technologies and procedures for the design. Otherwise, it's essentially impossible to identify them.*

Lun Ge: What you just mentioned above couldn't be more important. We have missed the contents and procedures that are indispensable. We have advocated the idea of preliminary planning since 2003. However, few hospitals responded. Currently, we work with the People's Hospital of Shandong Linyi District on its preliminary planning for a project approval, as well as the preparation for a project brief.

Jianhua Dong: Preliminary planning is a kind of brainstorm. Firstly, we make a thorough analysis of ourselves. After that, we figure out what the exact needs of a project would be, based on its practical situations, which naturally lead to an agreement on how to develop this project. Then, we can start designing a project brief and making a report on project feasibility. This is the way to ensure the maneuverability of a project planning, instead of being suffused with idle theorization.

Today, we have developed good foundations of preliminary planning research. Professor Lun Ge, you have accumulated a lot of theoretical achievements and experience. Your research findings have been applied in practice. There are excellent design teams emerging all over the country. This is a fairly good sign.

To sum up the first question, I think that the main problem in hospital development isn't regarding the process of medical services that have been raised to a high level, but the positioning and self-verification of the hospitals. *It's necessary for a hospital to make a thorough and indigenized analysis of itself, leading to a clear understanding of its future development* (Fig. 3.6).

Lun Ge: What do you think of the procedures of project bidding in this country?

Jianhua Dong: As to bidding for a construction project, people generally emphasize on building design and how the design is rendered. However, the most important thing is how the services and spatial functions are arranged in design. That's why we need to examine spatial design and service processing in project review. Only after that dare we evaluate the design. However, it's impossible to request every reviewer to have the ability to estimate design concepts regarding medical technologies and service processes. Often, they choose a design with beautiful renderings. It may turn out either there is no space for required medical services or it's difficult to process medical services there.

Although we have figured out how to make preliminary planning, we are still at a loss regarding what to do in early design. *Preliminary planning means more than consulting or verifying. It's in no way a formality. On the contrary, it's the core and the initial phase when we shall figure out who we are and what we want to be in the future.*

Project briefs have been known, but there is still a deficiency in developing a project brief. It's often out of touch with practical situations and thus couldn't be indigenized. In other words, the biggest problem is that project briefs generally lack maneuverability. It is the hospitals that provide project briefs, which are often intrinsically self-contradictory. Designers are thus afraid to proceed any further.

Fig. 3.6 Outpatient building of Beijing University First Hospital, Beijing

For instance, normally every department in a hospital writes a brief and then submit it to the hospital. However, it's nearly impossible for the hospital to make agreements on the needs of every department. It's just like a dish with all kinds of raw materials. How would this dish be cooked to cater all tastes? Each department has its own "taste". The hospital then leaves all these requests to designers to sort out. How is it possible? (Fig. 3.7).

Lun Ge: It's indeed of great significance to take all the needs into consideration and respond via spatial design.

Jianhua Dong: It's quite difficult as well. That's why it perplexes me a lot: brief development lacks maneuverability. Based on project briefs, the next steps are focused on project feasibility. Generally, two or three levels of project verification are required.

In project verification, the construction department and the planning department are required to participate. China has done a relatively good job in terms of project feasibility research. However, regarding POE, that's exactly what you professors advocate, it's still missing.

Those who truly engage in POE are the projects funded by the government. *The POE by government is majorly regarding the economic benefits of a project instead*

Fig. 3.7 Inpatient building of Beijing University First Hospital, Beijing

of social benefits. It concentrates on the balance between the input and the output of assets.

The situations of a project when it was initiated and when it was done for occupancy should be compared, in order to see whether or not a project's positioning is appropriate, etc.

I still remember that many years ago, I worked as a consultant for a private hospital in Shenzhen. They intended to build a general hospital with 1,500 beds. Based on a comprehensive analysis, I learned that the female workers in nearby factories would be the majority of their patients. Thus, I suggested the hospital to focus on the departments of gynecology, pediatrics, and obstetrics. Eventually, it has developed into a hospital with these special focuses. This is regarding project positioning.

Lun Ge: What you have engaged in is exactly what our hospitals need. What essentially differentiates your work from others is the maneuverability of project. As far as I know, you are the mentor of many architects. What is your advice to architects? Or is there any expectation on them?

Jianhua Dong: To architects, *the most important is to be insightful, distinctive, passionate, and creative*. It's kind of like writing: inspiration occurs all of a sudden. Dr. Xiqiu Huang is such an example in real life who often takes with him a note book. He told me that the note book was for writing down the ideas that suddenly popped in his mind for fear that he should forget. Only with passion and insights, will an architect develop distinctive ideas and then good works.

Lun Ge: What you said today is truly beneficial for architects. Thank you (Fig. 3.8).

Fig. 3.8 Jianhua Dong and Lun Ge

3.3 To Be a Good Translator and Interdisciplinary Facilitator—An Exclusive Interview with Rong Ma

Time: January 2015—reiterated in May 2023
Location: Nanjing Drum-tower Hospital, Nanjing

Rong Ma: Vice president of the Drum-tower Hospital in Nanjing (Fig. 3.9).

Key sentences:

1. *Successful hospital design and construction relies on neither the designer nor the hospital itself. It requires interdisciplinary cooperation.*
2. *Translators and interdisciplinary facilitators are important in hospital design and construction.*
3. *Upon designing a hospital, one must consider how the hospital systems could be run efficiently. So, the key to developing an efficient hospital lies in design.*

Fig. 3.9 Rong Ma

Lun Ge: Nanjing Drum-tower Hospital broke the traditional mode of hospital design in China. Cross China, many hospitals are designed in an anonymous and one-size-fits-all mode. With a new design concept, Nanjing Drum-tower Hospital has indeed brought surprises. After visiting it, I was very excited. It is truly a surprise to me that such a hospital could be built in China. Meanwhile, I admired the hospital president's courage to make such a decision to actualize the new design concept. As the construction manager, could you please talk about your thoughts on the design concept and the process of construction?

Rong Ma: Regarding the concept of hospital design, there are several aspects I'd like to talk about (Fig. 3.10). *First of all, hospital design is of the most intricate. It's impossible for any single architect to complete the design of a hospital even if he or she can design a five-star hotel or a commercial complex.* Secondly, during the whole process of hospital construction, communication between the hospital party and the architects is critical as well as difficult. Thirdly, there is only one set of codes regarding the construction of general hospitals, *National Construction Standards for General Hospitals*, which is quite brief and in need of specific indicators. Each hospital has its own peculiarities and regional characteristics. *The brief codes or general standards may not address the varied needs of individual hospitals.*

Regarding hospital management, based on my experience, it is focused on hospital construction and logistics. Firstly, we visited many domestic and foreign hospitals to learn their experiences. Impressive projects were found and each of them had a team of development. These teams respond to questions such as what the tasks and goals of a hospital are and what kind of buildings their hospitals need. They have thoughts on the planning of development. *Successful hospital design and construction relies on*

Fig. 3.10 Nanjing Drum-tower Hospital, Jiangsu Province

neither the designer nor the hospital itself. It requires interdisciplinary cooperation. Only by combining designers proposes and experts' comments, could a hospital develop a good plan. The key is that the hospital must have something in mind.

On the other hand, I pay special attention to two groups of people in our hospital: patients and physicians. Typically, it says that a hospital should be patient-centered. However, from my point of view, a hospital should put both groups in the first place. In our hospital, typically, an inpatient may stay for no more than 10 days. However, our physicians work here for 365 days a year. So, we need to pay special attention on their feedback regarding hospital environments. For example, we should be clear that there are cultural differences among the staff members. Exclusive areas should be provided for them to eat, talk with others, and rest. Providing these areas may help to reduce unnecessary contradictions between the staff and patients. More attention should be paid to the environmental experience and requirements of staff members.

Lun Ge: This is a good idea. How do you communicate and coordinate with designers or design institutes from foreign countries?

Rong Ma: In our project, the designer is a Swiss Chinese. Our team has frequent and smooth communication with him. His design concept seems quite modern but in fact traditional Chinese culture has been built inside. Our hospital adopted the design concept of a traditional Chinese courtyard, in which the atrium has nice natural lighting and ventilation, which is good to patients. *During our communication, we consider not only the aesthetics of the buildings but also the essence and needs of hospital development. This has gone through the entire process of our project.*

In order to ensure effective communication, a platform has been provided on which all the information and comments from individual departments can be gathered and transferred to the designer. He can then infuse the information in design. I participated in each of the discussions. We discussed hospital needs and service procedures. It works. *That is why I think translators and interdisciplinary facilitators are important in hospital design and construction.*

Secondly, it's necessary to collect comments from all parties throughout the project. When a problem arises, we should know who to discuss with and how to coordinate.

Thirdly, based on individual hospitals' characteristics, we should include local cultures in design consideration. Hospitals, patients, physicians, and nursing staff have their own needs. Hence, a platform for discussion and communication should be developed. I think "translators" and "interdisciplinary facilitators" are of particular importance in hospital design and construction. I collaborated with my assistants and project managers to build our communication platform. We conduct efficient communication with the designer.

During our communication with the Swiss designer, my mission is to let him understand the characteristics of Chinese hospitals and patients, in instead of asking him to draft floor plans in a rush at the beginning of a project (Fig. 3.11). We accompanied him to tour an outpatient department to help the designer understand why we need large waiting areas. We helped him to realize that the volume of patients in a Chinese hospital is different from that in a hospital abroad. Moreover, during the planning of a gastroscopy room, he provided design proposals which looked like those rooms in Germany. We selected one of them and then discussed with the designer regarding our preferences and needs in detail. In this way, we conducted efficient communication with the designer. The communication between hospital managers and designers is very important.

Although we communicated a lot with the oversea designer in terms of design ideology, we selected a local design institute to create and refine the construction documents. Regarding the structure design and the associated equipment, we gained safety permits from local government offices. However, this doesn't mean that oversea designers won't be involved in the discussion of design details.

During the entire process of design, the oversea designer participated in the concept design and preliminary planning, whereas a domestic design institute undertook the work of creating construction documents. However, the foreign designer was invited to participate in the discussion throughout the entire design. The designer was responsible to edit design drawings to meet user demands and sign design files. We conducted intensive discussions on design and adjusted the designs of operating

Fig. 3.11 Facade of Nanjing Drum-tower Hospital, Jiangsu Province

rooms, gastroscopy rooms, and magnetic resonance imaging rooms. After face-to-face discussions, design adjustments and modifications were then presented on the updated construction documents. During the discussions, the oversea designer put forward good opinions and relevant design experience from foreign countries for reference. After several rounds of discussion, we made final agreements on design and signed the design files. This method of collaboration ensures the integrity and unity of the whole project, from the early level of concept design to the later level of design implementation. Generally speaking, our cooperation is quite pleasant.

Lun Ge: Multidimensional communication is indeed necessary. Besides, could you please share more experience regarding construction management?

Rong Ma: *In terms of process management, we concentrate on functional demands and humanistic needs. It is important to find ways to standardize the management, simplify the process, and conduct project planning.*

First of all, from the perspective of systematisms, starting from the time of project bidding, the project brief must be clearly described with specific requirements. Generally speaking, how to conduct a spatial planning for humanistic care? To be more specific, what can we do for patients and our staff members? We should do systematic thinking at the macro level first and then solve the detailed problems one by one during building construction. Therefore, I think we should firstly clarify our goals as well as environmental plans at the very beginning. What we care about the most still is whether the interior environments can meet the needs of medical services and promote humanistic care. As a matter of fact, I'm not that concerned about the appearance of hospital buildings. Later in the communication with the designer, we were told that the appearance of our building has been consistent with its functions. We are fortunate that the designer's design concepts get along with our demands. The design is majorly focused on hospital functions and also informed with artistic ideas. It is a nice balance. Take the material selected for building appearance for example, the designer chose frosted glass but many of our hospital representatives preferred clear glass. The designer informed us that frosted glass could diffuse light and thus more comfortable. We adopted his design to enhance natural lighting and promote environmental comfort. It turns out that such design meets the functional demands for medical care very well.

First, we should view general hospital planning from a perspective of systematisms. For example, regarding the spatial and functional programming, we should set up goals and general plans at the beginning.

What I care the most is whether the design of a hospital environment is corresponding to its function, whether the environment could promote humanistic care for patients and staff. I don't really care about the building appearance. Fortunately, the appearance is consistent with its function. The inclusive and artistic design coincides with the building's functional needs.

For example, the designer chose frosted glass which diffused light, instead of clear glass preferred by hospital representatives. The frosted glass could diffuse light and thus may make the patients there feel more comfortable and bring psychological pleasure.

Secondly, in architectural design, with the routine procedures of hospital visiting in mind, we reserved some spaces for flexibility. Specifically, in the waiting areas for outpatients, clear signages and instructions to outpatients similar to those provided in airports for passengers should be available to reduce delay and congestion associated with bad way finding. This can mitigate the situation of lingering and reduce congestion. It provides convenience to patients and their families. To facilitate medicine pickup, the pharmacy areas should be equipped with intelligent systems providing row numbers, just like banks do, to improve the efficiency and accuracy. We may borrow the methods of number assigning from banks. It helps to control the volume of visitors in the areas. On the other hand, it promotes the informatization of services and the efficiency and accuracy of filling prescriptions. All of our medicine warehouses are located on the first floor underground. Medicines are transferred by automatic machines to the pharmacies on each floor. Good design and planning provide convenience in practice but no demands for renovation or transformation. That's why I insist that designers should communicate with hospital professionals regarding hospital management and functional demands at the beginning of design.

All of our pharmacies are located in the basement and equipped with automatic dispensing machines. *Upon designing a hospital, one must consider how the hospital systems could be run efficiently. So, the key to developing an efficient hospital lies in design.*

Furthermore, of a hospital, the interior decoration is important and requires uniformed styles. The interior design of Drum-tower Hospital is distinctive as it reflects not only humanistic care but also integrates with art and religions. This hospital is not simply a hospital, but more of a museum or an art gallery. In fact, we have given relatively more flexibility to the designer team in this respect, because that's their profession and what they're good at. In the process of construction, the designers set up specific requirements for technologies, colors, and etc. For example, a mild wall-color has been suggested. Most of the lighting colors in rooms are mild as well, which make people feel warm.

In addition, in terms of construction quality control, on the basis of holding these key points, we should pay attention to effective and timely communications and discussions, so as to solve problems efficiently.

Lun Ge: I agree. Management should be concentrated on the key points. In my opinion, first, project programming is required before collecting a project approval. Second, if approved, a project brief should be required. Third, building design should integrated with medical programming. What are your comments on these points?

Rong Ma: *I think conducting consultation in early design is critical and should be done by professionals.* On the other, there should be certain methodologies and specific codes for managing hospital development. Furthermore, we should set up education programs to train "translators" who can cultivate ideologies and awareness systematically. I*n my opinion, the processes of project inspection and evaluation are of high significance.* Once a hospital was built, it should be built under the standards set by the Ministry of Housing and Urban–Rural Development. *If enough attention were paid on post occupancy evaluation, it would have* pre-emptive effects

Fig. 3.12 Rong Ma and Lun Ge

on encouraging people to consider hospital development in a more forward-looking and systematic manner. *Hospital design isn't a one-way process but a circular chain of vivacity* (Fig. 3.12).

Lun Ge: It is my honor to talk with you today. Thanks again for sharing!

3.4 Tackling Architectural Challenges with "Anatomy"—An Exclusive Interview with Xin Ni

Time: December 2014—reiterated in May 2023
Location: The president's office in Beijing Children's Hospital, Beijing

Xin Ni: Chief Physician, Professor, Doctoral Supervisor; Director of National Center for Children's Health; President of Beijing Children's Hospital affiliated with Capital Medical University; President of Beijing Pediatric Research Institute; Dean of the School of Pediatrics at Capital Medical University (Fig. 3.13).

Key sentences:

1. *Being a medical doctor, I am specialized in human anatomy. Similarly, you architects are also "medical doctors", who work on the anatomy of buildings. My work on human anatomy may be more complicated than yours. Now that I've done perfectly in such a delicate domain of human anatomy, the 'anatomy of medical*

Fig. 3.13 Xin Ni

architecture' wouldn't be intractable to me. So, I hope you can trust me and let us work together to get the project done successfully.

2. *Never undervalue the importance of hospital logistics and the logistics personnels. These support a hospital' infrastructure. A hospital cannot function well without a safe built environment. We should learn from all aspects to gain valuable experiences in practice.*
3. *No matter about its type, a hospital should aim at completing diagnosis and treatment services through the shortest pathway and in the shortest time.*

Lun Ge: Hello, President Ni, you are a medical expert. How do you enter the field of hospital construction? Could you share your thoughts and experience regarding hospital construction?

Xin Ni: Actually, it was at Beijing Tongren Hospital where I participated in hospital construction at the first time. After my graduation from Harbin Medical University in 1991, I was assigned to Beijing Tongren Hospital. Later, I got a Master's degree in Medicine and Management from Capital Medical University and went back to Harbin Medical University for a Doctorate of Public Health. It was in 2002 when Beijing Tongren Hospital planned to build a branch in Yizhuang, Beijing (Fig. 3.14). I was the director of the project office and in charge of the entire process of building design and construction, from approval application to construction management.

This project was processed quite rapidly. In the March of 2001, we started working on the application for a project approval, which was received one year later in March, 2002. Then, we visited many hospitals across the country to investigate hospital design. The construction work was initiated on the 18th of April in the same year and the new hospital was officially opened on the 18th of May, 2004. As far as I know, in the construction history of China, no hospital has ever been built faster than that. We received a "Great Wall Cup" award for the structural design and a silver award of "Excellent Architecture in China" for the entire project, which was ranked 5th of the ten silver awardees. Only one gold award was announced at that year and it was for a national project.

There are several reasons for this project to make such an achievement. It was called as "a modern hospital breaking out of a land of sorghum." First of all, the

Fig. 3.14 Beijing Tongren Hospital in Yizhuang District, Beijing

site was 'brown' and had no residents so the construction work could be conducted for 24 h per day. Secondly, we proposed a design philosophy that is "keeping pace with the time for at least ten years". In terms of design, a courtyard is set in the outpatient hall to make it an open space. A patio is included in the hall to facilitate access to it. This kind of design was quite uncommon at that time. In addition, we saved half of the 200-mu (about 133,333 m^2) available land for future use. Being "moderately advanced" is important. Thirdly, since this project was funded by self-raised funds instead of fiscal supports from the government, we tried hard to cut down the costs. Upon project completion, the total cost was about 3000 CNY per square meter whereas the average unit cost estimated by the City of Beijing was 6000 CNY per square meter.

In our first meeting on project supervision, both the construction supervision company and material suppliers participated. I said "Being a doctor, I am an expert of human anatomy. Similarly, builders are 'doctors' who have expertise in building anatomy. My work regarding human anatomy may be more complicated than the building anatomy. Now that I've done perfectly in such a delicate domain, the 'anatomy of medical architecture' would not be intractable to me. So I hope you can trust me and let us work together to get the project done successfully." Since then, our collaboration has been run very well. It made a deep impression on me.

At the beginning, I knew nothing about architecture. It took me about three months to learn the basic concepts and rules. Our hospital set up a lot of requirements to ensure construction quality, especially regarding the implementation of concealed tasks. It was a learning by doing. My architectural experience accumulated in this project has been of great significance to me. At that time, I was the vice president in charge of medical services but earned a lot of experience beyond medical services, such as logistics, construction, and hospital operation. In a word, it's wrong to think that logistics plays a minor role in a hospital and medical services always have the priority. *Actually, logistics is an indispensable part in hospital management. Hospitals couldn't run without good logistics and safe environments.*

Lun Ge: It is such a vivid figurative. We often compare building systems with the systems of human bodies as well. Any other projects? How about the Children's Hospital?

Xin Ni: At that time, Beijing Tongren Hospital established the Tongren Medical Group, which then built two branches in Nanjing and Kunming known as Nanjing Tongren Hospital and Kunming Tongren Hospital. I worked concurrently in the Tongren Medical Group as the Vice President and was responsible for both medical services and construction. As the representative of Beijing Tongren Hospital, I participated in the entire process of the new branches' development, from pre-design to building construction.

Regarding the Tongren Hospital in Yizhuang, Beijing, it was proposed to be an 18-story building, in order to be the tallest building and a landmark in the Yizhuang development area. I opposed to this idea because being a landmark didn't necessarily mean to be the tallest building. What a hospital should pursue is efficiency: physicians should be efficient in-service provision and so do the patients in receiving services. A long time waiting in hospital generally has a negative impact on patient mood. To promote service efficiency in a hospital, in my opinion, the horizontal functional flows work better than the vertical ones. Moreover, high-rise buildings require large investments to install elevators. Eventually, the Yizhuang branch was built as a nine-story. The key is to focus on the needs of medical services instead of building appearance.

Lun Ge: What you just mentioned is what we have drummed up for hospital design. As a medical expert, it's extraordinary of you to have such a profound understanding of hospital design and construction. We would like to learn your insights regarding the coordination and cooperation with other majors during the process of hospital construction. What are the aspects in need of improvements? Are there any serious problems that reduce the quality of project?

Xin Ni: Regarding design, I think we should choose a design institute with rich experience regarding hospital design. To build a hospital, I could only provide some rough ideas, such as building functions, sizes, and symbols. It would be helpful if the design team had experience handling the needs of a hospital, such as service procedures, functions, volumes, and requirements from clients. Regarding this, I suggested the government to designate several design institutes with expertise in hospital design for hospital projects. The more focused a design institute is, the more experienced it would be.

How much on earth does a hospital president know about building design and construction? Regarding building and construction, what hospital presidents know may be very limited since they spend the majority of their time in scientific research and academic promotion. The "professionalization of hospital presidents" has been advocated. This leads to the philosophy of using professional managers as the presidents. Without a strong medical background, the presidents cannot earn sufficient respect in their hospitals. Meanwhile, they also need to be professional managers. What shall we do in this case? We need an interdisciplinary team in hospital design to assist the president in analyzing the hospital's needs, positioning, and specific functions. *To summarize, in order to design a hospital well, we need the professionals of design, policy making and implementation to collaborate all together*.

Lun Ge: Here is the second question. In comparison with a general hospital, how to differentiate the characteristics of a children's hospital? Could you please share your thoughts?

Xin Ni: Nowadays, it's widely advocated that a hospital should be differentiated by the tone and scent of its environments, and the looking of its staff members. For instance, the physicians and nursing staff in our hospital wear the suits decorated with printed flowers, birds, and rabbits instead of the traditional uniforms in blank white so that the children there may not experience a feeling of hospital. To cure diseases, 30% of the work is relied on treatment whereas the remaining 70% is on rehabilitation. A children's hospital should address the needs of children.

Firstly, the environment in a children's hospital should be bright and pleasant, promote care and joyfulness, and help to eliminate anxiety and fear (Fig. 3.15). Typically, children seeing a doctor are accompanied by their families, whose feelings should also be considered in hospital design. More importantly, children are vulnerable so we should pay extra attention to the details of the design. All sharp edges should be well coped. The materials used in the wards should be soft and the floor should be slip-resisting. These are suggested features for a children's hospital.

Fig. 3.15 Beijing Children's Hospital, Beijing

Lun Ge: You talked about the images of a hospital. How about the process of medical services?

Xin Ni: *No matter about its type, a hospital should aim at completing diagnosis and treatment services through the shortest pathway and in the shortest time.* This is required for being a qualified hospital. When I joined the Beijing Children's Hospital, the media introduced me as "a new president who improved the service procedures and made an average of 24-min reduction of waiting time for outpatients with fever". Prior to that, regarding a child with fever, body temperature is measured during hospital registration. If the reading were higher than 38.5 °C, immediate services would be provided by nurses to lower the temperature to avoid febrile convulsions before the patient joined the waiting line. Generally, when the patient arrives at a doctors' office, before providing any further treatment, the doctor needs to see the results of blood counts and C-reactive protein examination. It means the patient needs to go through another line of waiting to take the exams before returning to the doctor. So, the patient had to wait twice for the final results. To solve this problem, I changed the service procedure to make the two tests be ordered right after registration or if the temperature reading were over 38.5 °C, so the patient could have the exam results ready at the time seeing a doctor. As a result of change, the waiting time was cut by 24 min on average. To the patient, this means far more than just the time saved.

This case of shortened patient wait times is regarding the design of functional flows. Similar cases happened regarding hospital management and architectural design. Management and architecture are closely related. Hospital operation can be optimized with the integration of management and design.

Actually, this identifies another problem of hospital management, which needs more discussion. A lot of decisions regarding hospital design and construction are made subjectively by the presidents or department heads, rather than scientific research. Changes in project planning are frequently associated with the changes of department heads. To solve the problem, the standards and procedures of design and planning should be kept at all the time regardless who the department head were. The key is that the leaders should follow the plan but no others.

During hospital design and construction, verification is usually insufficient. Many things are in a manner of subjection instead of institutionalization. Take the development of a hospital department as an example, no matter who the department head were, the procedures of hospital development should be stipulated. Project modification at the department level should follow the overall planning. No matter who the leader was, things should be done according to the planning. This is a core rule of hospital development.

Lun Ge: The third question is regarding the phenomenon showing whether or not we can have a high-quality hospital is by chance. Once a responsible president and designers joined the project, the quality would be ensured. Otherwise, it's the other way around. Given the massive government investments in hospital projects, this phenomenon is really pathetic. So, what do you think about the pre-design and planning of a hospital?

Xin Ni: Preliminary consultation should be included in early design. It helps make a foundation to propose the project and may take 3–6 months to clarify everything before submitting an application for project approval. During the time of developing a project proposal, things are basically determined upon consultation. During 3–6 months of consultation, we figure out project needs and what we truly want to build. Based on that, we develop the proposal and submit it for reviewing. However, a few people are aware of the importance of preliminary consultation in early design and many start consulting after receiving a project approval. At that time, they also work on fund raising and construction committee development so the time spent for consultation is very limited. *Therefore, it is essential to conduct preliminary consultation at the very beginning of a hospital project, which has been adopted by the Health Bureau as part of their project management.*

Lun Ge: Actually, we share the same thoughts regarding hospital construction processes. We have initiated a program on post occupancy evaluation—POE. We brought forward the "SHAPE" program to conduct the POE of built hospitals.

Xin Ni: This would be very helpful. How could we make improvements without POE? The PDCA (Plan-Do-Check-Act) cycle in our management would fail without having the evaluation as part of itself. We need to know whether or not the project has been built as expected. Hopefully, it can provide a reference for future projects.

Lun Ge: Many hospital presidents are hesitating to conduct POE since they worry that possible mistakes identified in the evaluation would be used to blame them.

Xin Ni: POE is not about identifying mistakes but revealing functional flaws associated with improper design, and thus calls for designers' attention. Nevertheless, *the institutionalization of POE would make things easier.*

Lun Ge: What you said hits the point. A cycle stops running when the chain breaks. We don't know how the government funding has been possibly wasted and where the problem is. We have been advocating the idea of "evidence-based design" that suggests the facts to speak for themselves. From this standpoint, hospital construction in our country is in great need of POE. I'm happy that you agree with us. Lastly, what are the aspects do you think we should control in order to ensure the quality of hospital development?

Xin Ni: *How to control the quality of hospital construction, including the "concealed projects" that couldn't be seen from the outside? Firstly, integrity is of top importance to both the constructors and supervisors.* Honesty comes first no matter in terms of the construction or supervision. Secondly, we should minimize subjectivity and reduce the needs for constant supervision. Hence, there should be some matured management standards for people to follow. Thirdly, we should entrust projects to experienced professionals. Moreover, the government should set up standards to clarify a contractor's qualification.

Besides, there is an issue of expenditures. Why do certain people say that the "concealed projects" could save money? Every enterprise seeks profits, but how? A list of materials suggested for construction should be developed. Once we make decisions on materials, relevant personnel costs should go with national standards. Negotiation is allowed. However, the main materials shall never be changed. This is one of the ways to guarantee the quality of project.

Expenditure is another issue. Some company would take advantage of the "concealed projects" to make extra profits, which is certainly not right. To be specific, all projects of over 300,000 CNY should be audited, with the audit fee being part of the difference between the costs reported by the contractor and the costs approved by the auditor. Let's say, the contractor asks for 500,000 CNY whereas the audited costs would be 450,000 CNY. The difference or the saving is 50,000 CNY so the audit fee should be a percentage of the 50,000 CNY. To increase the fee, auditors would cut the costs as much as possible. In order to make profits, the contractor would try to save money from the "concealed projects" as much as possible and thus unavoidably results in a seemingly good but actually substandard concealed project. That's why I set up specific requirements regarding the essential materials applied in our projects whereas the contractor may make decisions on other materials. In this way, we are able to guarantee the project quality while leaving room for the contractors to make profits (Fig. 3.16).

Lun Ge: All you mentioned above hits the spots. Thanks for your precious time!

Fig. 3.16 Xin Ni and Lun Ge

3.5 Building Hospitals: Blessings Bestowed by Our Practice—An Exclusive Interview with Chongde Shen

Time: January 2015—reiterated in May 2023
Location: The office of Chongde Shen in Wuxi People's Hospital, Wuxi

Chongde Shen: Vice President of Wuxi People's Hospital, Supervisor of Masters' and Doctor's programs. Chief Member of the Standards Professional Committee of China Health Information Association; Vice Chairman of the Information Statistics Professional Committee of China Hospital Association (Fig. 3.17).

Key sentences:

1. *In many projects, on the first day of occupancy, building renovation starts. These problems result from deficient systematisms and professionality.*
2. *Traditional Chinese Medicine hospitals promote Chinese cultures as well as Chinese medical cultures. TCM is the quintessence of Chinese cultures and offers the most concentrated expression of Chinese philosophy.*
3. *Although we work in the field of hospital logistics and don't directly face the patients, we actually indirectly provide services to them and should be dedicated to a great love and with valuable recognition of the job that is to support patients and their families, and also to let the employees have a correct understanding of themselves. We help them to understand the uniqueness of their work. It promotes universal love.*

Lun Ge: It is nice to meet you, President Shen. As a medical expert, you are also an expert in the field of hospital management and construction. I joined your presentations at the China Hospital Construction Conference and was deeply impressed. Could you share your thoughts and experience working on hospital construction projects?

Fig. 3.17 Chongde Shen

Chongde Shen: *I think there are two big aspects in the field of hospital design and construction: systematism and professionality.* The systematism includes two issues. The first one is regarding the procedures. It is focused on the process of project development, from planning to acceptance inspection and maintenance. What should be done in each phase of the development? Many project participants don't understand their project even after it has been completed. Certain real estate companies such as Vanke (a famous real estate company in China) do very well in systematisms. Let us take Vanke's project application as an example. Their application files include many pages of functional flow charts showing expected tasks in each stage of the project development. Based on a high-quality systematic plan, the proposed project would be developed in a high speed and with high quality. In contrast, a lack of systematisms may lead to a project shut-down. However, some managers have no idea on the importance of project verification and also don't know how to identify professionals who can conduct the verification.

The second issue of systematism refers to the lack of a systematic understanding of one's project, including its inherent capabilities, layouts, business processes, future operation, management, and service modes, etc. Without a strategic perspective and a systematical consideration integrating individual factors of hospital development, such as medical treatment, scientific research, prevention, health care, and rehabilitation, the development would be a journey without destination. We reiterate the importance of disease prevention, medical care, and rehabilitation. *The point is that we need to provide medical services within the framework of our projects and give priority to selected key factors.* In terms of medical care, we should understand what the key disciplines are, what the priorities are, what kind of technologies should be used, how many devices are needed, and the appropriate scales of them, etc. All of these are involved in the systematism of a hospital's design and construction. Therefore, we should systematically figure out these questions in order to build a high-quality hospital.

The second aspect is regarding professionality. Systematism is reflected in the cognition of professional knowledge. An agent construction company may say "I have built so many buildings. How come I don't know how to build a hospital?" Likewise, many design institutes also think "there are norms for building design. How difficult could the construction of a hospital be?" Many hospital constructors would say "we have been building hospitals as always." As a matter of fact, they make mistakes in cognition. *Professionality is shown at the key points of a project.* For example, in a HVAC (Heating, Ventilation, and Air Conditioning) design, we typically focus on the cooling and heating whereas little attention has been paid to the ventilation. Further, energy-saving strategies require an in-depth infection control that has special standards for ventilation zoning. In addition, professional management is required. *How to grasp the key steps and procedures of project development is critical.* Here is the current situation: we are in need for professional plans focusing on material examination, quality evaluation, and project inspection, etc. Many supervision schemes are deficient in professionality to some degree and need to be more concentrated on the targeted items. Moreover, the overall development of hospital design and construction needs attention. *We haven't provided any holistic training on construction management and professional cognition regarding hospital development.*

As mentioned earlier, the issues of systematism and professionality have been shown in practice. Some hospitals might have no idea about building design and construction. Sometimes, when a construction work had just been done, including walls and pre-installed items, building renovation would then be started. Frequently, 40–70% of the construction work is demolished. Of the embeds, about 40–50% is abandoned. The reasons are associated with the standards for construction document reviews. The standards are regarding general construction and don't pay special attention to medical technology and its fast development. In consequence, although the document passes the review, it doesn't meet the needs for hospital development. Many hospitals implement and refine a design at the same time and modify the construction while building it.

A large number of discrepancies show up during the phase of interior design and decoration. *In many projects, on the first day of occupancy, building renovation starts. These problems result from deficient systematisms and professionality.*

Lun Ge: Your enquiring mind and profound understanding are very impressive. Could you share experiences regarding hospital development? What is the part that makes a deep impression on you?

Chongde Shen: I'd like to explain it from the perspective of a Buddhist. A lot of people are unaware of the fact that the meaning of hospital development is to help whoever needs help. Hospital developers intend to create good environments for the helpers such as healthcare professionals to work with and pursue career goals (Fig. 3.18). Isn't this more worthwhile than traditional charities? If the overall planning or systematism of a hospital had issues, the hospital should invite professional institutions and experts to search for solutions. On the other hand, hospital managers should conduct self-teaching to learn knowledge on design and construction. No one is born to be a professional. A hospital manager should have the characteristics of a design professional and respect hospital development with a sense of historical responsibility. The functionality of a hospital is foremost, after which comes the normativity and then the degree of aesthetics and comfort. We should never trade functionality for appearance. In many projects, we see construction agents push forward the process of construction at the cost of buildings' function and opportunities for future development. This violates the law of objective development and is regrettable and distressing. We should bear in mind a sense of responsibility and work with professional cognition and attitudes, in order to promote hospital development.

In addition, we should implement management ideologies into construction projects. Regarding the management of a large hospital, the sage leads, the wise advises, and the sweaty toils. What we need to do is to define the ideas, mechanisms, processes, supervise and provide trainings to the staff, rather than doing it in person.

Ideology remains the key. For instance, we should consider how to help staff members realize their value through logistics management. *Although we work in the field of hospital logistics and don't directly face the patients, we actually indirectly provide services to them and should be dedicated to a great love and with valuable*

Fig. 3.18 Aerial view of Wuxi People's Hospital, Jiangsu Province

recognition of the job that is to support patients and their families, and also to let the employees have a correct understanding of themselves. We help them to understand the uniqueness of their work. It promotes universal love. This is what I have emphasized in our staff training.

Lun Ge: Your training is based on a value system of humanity, not just techniques. It has a far-reaching impact on staff members. You have participated in the construction of multiple traditional Chinese medicine (TCM) hospitals. Could you please talk about the particularities of TCM hospital construction?

Chongde Shen: There are two particularities: hospital functionality and the emphasis on humanistic cultures. In TCM hospitals, there are two modes of business. One is in a completely traditional format that has been widely adopted by small TCM clinics. Differently, the new mode has modern technologies integrated with TCM characteristics. In the new, the hospital is a general hospital in the first place and then provides TCM services. Because of its functionality, the spatial layout of a TCM hospital differs greatly from that of a general hospital. In addition to medical functions, a TCM hospital should have rich cultural characteristics as well.

Regarding hospital functions at a macro level, the classification of services provided in a TCM hospital is different from those in a general hospital. Traditional Chinese medicine advocates prevention, such as "to prevent before disease onset and after the recovery." It leads to the special requirements for TCM wards. Further, rehabilitation services in a TCM hospital require attention. Rehabilitation helps patients such as those with hemiplegia to recover, return to the society, and revitalize their families. There are huge demands for rehabilitation care and TCM meets the needs. Another particularity of TCM hospitals lies in their pharmacies. For instance, TCM hospitals require certain medical processes such as TCM preparation and decoction, which are unavailable in western medicine hospitals. Some western medicine hospitals may have small TCM pharmacies. Moreover, the spaces for acupuncture, massage, pharmacy, and medicine storage in TCM hospitals are unique. TCM hospitals should emphasize on their particularities, such as TCM prevention and rehabilitation.

Given the particularities of TCM, the layout of a TCM outpatient department should be able to accommodate a large volume of outpatients. Most TCM outpatients have chronic diseases and need to visit hospitals for many times. The typical duration of their hospital visits is relatively longer than that of the outpatients in a general hospital. Therefore, it is inappropriate to use the standards for general hospitals to design the layout of a TCM outpatient department.

Generally, to accommodate the same number of patient beds, a TCM hospital requires more floor areas than a western medicine hospital. Why? A TCM hospital may not need rooms for medical testing or experimentation but requires a variety of spaces for acupuncture, massage, and rehabilitation. Moreover, pharmacies in a TCM hospital are typically larger than those in a western medicine hospital. Special characteristics of TCM are not only shown in the field of medicine but also regarding the room layouts and service procedures. For instance, waiting areas are critical in TCM pharmacies since it takes time to make TCM. It would be over-crowded by patients and families queuing up for medicine if no waiting areas in the pharmacies.

The second particularity of TCM hospitals is regarding humanistic Chinese cultures. Cultural characteristics can be presented through architectural design. It does not mean that glazed tiles and cornices are required in TCM hospitals. However, it is preferred that the design of a TCM hospital embodies some TCM features and regional characteristics. Once entering the hospital, people would instantly realize that it is a TCM hospital instead of any other types of hospitals. Actually, in regard to this, the National Administration of TCM has specific requirements.

TCM hospitals promote Chinese cultures as well as Chinese medical cultures. TCM is the quintessence of Chinese cultures and offers the most concentrated expression of Chinese philosophy. Humanistic culture is mainly shown when people "experience" it. How to build humanistic environments in a TCM hospital? It is what we should think about (Fig. 3.19).

Lun Ge: It's a pity that limited explorations and research have been done regarding TCM hospital design and construction. TCM particularities should be promoted. What are your experience and advice regarding the procedures of hospital development?

Chongde Shen: From the perspective of hospital management, preliminary planning is critical in the entire process of hospital development. Although the analyses of SWOT, value chain, competitiveness, etc. are not necessarily a must, it is always required to figure out what a hospital is going to do in the future and how to make it come true. Furthermore, we need to consider the strategies of hospital development in a long run as well as in a short run. It's suggested that a hospital should invite a third party to make professional evaluations on hospital development and develop a project brief based on marketing and development needs. Regarding hospital construction, firstly, there should be a medium-term or long-term plan. Then, specific construction goals, principles, and development procedures should be established for each phase of

Fig. 3.19 Wuxi People's Hospital, Jiangsu Province

the development. A project brief of thorough and intensive development tasks should be formulated. We shall neither shackle the mind of architects with our thoughts on minor details, nor be slapdash to the extent of forgetting what we want to build in general.

As far as I'm concerned, a project brief should be carefully developed step by step. Firstly, we make a brief for the phase of design development. The second step is focused on medical services, a technical planning, and a special project planning. This method helps designers to handle a large amount of information and design tasks. Importantly, proper pre-design planning is critical since the absence of it would result in a huge waste in construction.

There were two deficiencies in the early stages of our construction management. The first is the lack of professional and systematic research on key issues. We should invite third-party experts and professional institutions to help us make more scientific decisions on project planning. The second lies in the insufficient understanding regarding the integrality of construction technologies and procedures, which is the core of construction management.

The first deficiency involves cognition whereas the second influences practice and has a direct impact on the quality of construction. As a result, I believe that there are gaps to some extent in these fields. From the perspective of a preliminary planning, the preparation of a project brief and the design of service flows are particularly needed. It is of guiding significance to address these gaps in hospital industry and promote the virtuous cycles of hospital development.

Lun Ge: Thanks very much for your wonderful sharing (Fig. 3.20).

Fig. 3.20 Chongde Shen and Lun Ge

3.6 Solving Professional Problems Professionally—An Exclusive Interview with Tielin Wang

Time: January 2015—reiterated in May 2023
Location: Tianjin

Tielin Wang: President of Hainan Cancer Hospital, former Vice President of TEDA International Cardiovascular Hospital in Tianjin, former President of Tianjin TEDA Hospital, Member of the hospital construction expert panel of the Health Ministry of China, President and Member of the Building Committee of Chinese Hospital Association, Drafter of the *Design Specifications for General Hospital Buildings*, *Construction Standards for General Hospitals*, *Construction Standards for the Clean Operating Departments of General Hospitals*, and *Technical Specifications for the Clean Operating Departments of General Hospitals*, Editor of the *Chinese Hospital Management Encyclopedia*, *Guidelines on Hospital Surgical Department Construction* (Fig. 3.21).

Key sentences:

1. *A hospital should be an organic system. The deeper we investigate this system, the more we recognize the importance of utilizing integrated spaces. Every square meter should be used appropriately.*

Fig. 3.21 Tielin Wang

2. *We lack medical planners and need professional designs, construction, and consultancy. A project review would be meaningful if it improved the project and guided its future development.*
3. *It wouldn't be a matter if all my words got deleted from this book but saving a single sentence saying "Space program and function program are the foundation of hospital design".*

Lun Ge: President Wang, it is glad to see you again. It has been eight years since our collaboration on the project of Tianjin TEDA Hospital. I learned a lot from you. Besides the TEDA Hospital, you participated in several other hospital projects and must have gained a lot of new thoughts and experiences regarding hospital construction. Our research is focused on the procedures of hospital management and construction, which is of great importance in China. Could you talk about it?

Tielin Wang: First of all, I think it is very important and necessary to ensure the quality of work at each level of a project. *Of a hospital project, the preliminary planning and schematic design directly influence the quality of design and budget making.* In early design, it's critical to develop a systematical framework of project development and clarify the specific requirements by level. The project can then be constantly refined. Please allow professionals to refine design drawings. A systematic design framework shouldn't be in too much detail. *Architects may set up the general "requirements" and leave the rest to be refined by professionals majoring in different fields.*

Regarding the confirmation of project completeness, it should include a medical inspection and an architectural inspection. People typically need time to adapt to the new buildings. It takes time. We shall predict in advance what kind of problems could arise in a project inspection. It's like setting a goal that can be investigated in detail in the inspection. A confirmation inspection should be conducted independently and as a separated phase of project. The processes of a hospital project include three stages: the early design stage, the medium-term stage including the design of construction documents and the construction, and the stage of inspection. With this concept of three stages, it is easy to distinguish, understand, and discuss project development.

There are two main contradictions: *Firstly, where should the preliminary planning end? Secondly, what should we inspect at the stage of inspection?* These problems in process should be solved in the process. Some of the standards are confusing. Let's take the Evaluation Standards of Green Buildings as an example. It is required to see the result of building performance after two years of occupancy. That is to say, the factors of building management and operation need to be considered in evaluation. However, these factors vary by geographic location and socioeconomic level so it is difficult to unify the standards nationwide. The formulation of standards should be focused on building design and construction rather than administration. At present, issues of architecture and administration have been mixed up and architecture needs to address administrative problems.

Lun Ge: Do you think it's necessary to conduct evaluations on a hospital project, in terms of functional design and space utilization, after a two-year occupancy?

Tielin Wang: There is a question of feasibility. We should consider the payment for post occupancy evaluation (POE) and what a hospital could get from the evaluation. It should make sense to the hospital. If the hospital were a stuff for sale, as the owner, I would like to have the evaluation so that I can sell it for a good price.

To a hospital manager, it would be all right if the evaluation could be used to supervise contractors or improve hospital performance or effectiveness, otherwise what is the purpose of evaluating? If asked me or a hospital project, to take an evaluation after two or three years of occupancy, no matter it is about my project performance or influence, as far as the evaluation results are good to me, I would be willing to evaluate it. Nowadays, everyone wants to win a National Excellence Award or a Luban Award. It firstly represents a high level of project quality and secondly represents an excellent contribution and performance of the builders, so both the construction companies and the project owners are willing to take these awards.

In addition, *something is wrong with our construction modes. T*he problem is regarding the general planning instead of specific investments or technologies. So far, we lack specialized majors as well as professional service providers in hospital design, engineering, and consultancy. POE would only be meaningful if it actually facilitated hospital development, or played a role in promoting and guiding the development.

Another issue is that we lack a system of general contracting, although it has been carried out by the Ministry of Construction for years. Why? Because the influences by hospitals is very strong in hospital projects and their interests always dominate the project development during the whole process. Only based on a completed project supervision and quality control, the general contractor systems can be established.

Lun Ge: You mentioned "General Contractor Systems" whereas there are ongoing "Agent Construction Systems". What is your opinion on the Agent Construction Systems?

Tielin Wang: The presentation of "Agent Construction Systems" means that the government is worried about the quality of hospital construction and thus requires the systems. However, the current systems fail to solve the dominant problems in construction management. The in-control power behind the current "agent construction systems" is not from project management companies, but the shadow of clients. Clients control the systems in the back. It is not about whether or not the client is good but professional enough to control.

In the absence of a precondition, I personally oppose to the "agent construction systems" since its application would add an additional 'Director.' The two directors may have some conflicts and their distribution of benefits could be an issue. So, the number of directors can only be one.

I think the majority of a construction project should be clearly identified and the legal representative should be in place. The major part such as a hospital should bear the legal responsibility of construction. Only when the legal responsibility was identified could the project be well developed. If in need of professionals in construction and design, the hospital may hire these professionals to solve professional problems. Now, specialized institutions or agents are required but we are short of them. We should emphasize on cultivating such specialized institutions but this may require some adjustments to the current modes of construction.

Lun Ge: In your opinion, what is a hospital evaluation? How to specify the items of evaluation?

Tielin Wang: More specifically, it refers to the evaluation of a hospital's functional abilities to provide medical services. On one hand, it is focused on functions. On the other, it aims at evaluating the services associated with specific medical care.

For example, in order to build a linear accelerator room, we shall consider how much its power capacity is, whether or not the functional layout is reasonable, whether it meets the standards for radiation protection, how many accelerators the hospital should have, what type of equipment should be installed, what kind of departments should be built, and whether or not this planning conforms to the positioning and orientation of our hospital development. Upon the request of the Health Bureau to build a hospital, first of all, we should know whether this hospital is included in the regional planning, the hospital type that is proposed in the planning, the hospital level that is needed in the planning, and how many departments this hospital needs according to its positioning. These departments would define the service framework of this hospital. Based on this framework and its workloads, we can then estimate the number of accelerators needed in the radiology department and the scale of its laboratory department, and ensure that these departments meet the corresponding needs for medical services. For example, evaluating the size and functional layout of a laboratory department would lead to a series of questions on medical care.

Lun Ge: In addition, preliminary planning is very important before collecting a project approval. What do you think about that?

Tielin Wang: In the book *'Hospital Management—Construction'*, I edited the Chapter of Report on Project Feasibility, which elaborated the work that should be done at a certain stage and the key points. No matter what this work is called, a proposal of preliminary planning or a report on medical processes, in fact, the contents are similar. The core of a project brief is the report on technical processes for medical services.

If the tasks of design development were divided into an early design and a report on technical processes plus a project brief, the key points of project quality control would be identified when the design development were done, and the rest of design can then be processed smoothly.

Therefore, a project brief must be done before initiating the design development, and the report on project feasibility should be drafted after completing a design development. When I wrote the Chapter of Preliminary Planning for the book, I suggested that a preliminary design or early design should be included in the stage of design development.

Once you moved on to the stage of design development, everything regarding the project has been included in your framework. At that time, it would be difficult to make a change. In fact, the files requested by the NDRC (National Development and Reform Commission) are at the level of schematic design. Of them, the first thing is regarding architectural planning and the second is a schematic design of individual systems included in the project. At this level, it would be good enough for project development if the schematic design is clearly presented. By doing this, we just make an agreement with the NDRC, whereas the document of design must be approved by the design institute and stamped with seals.

Lun Ge: Then, the tasks of a project at an early stage are clarified. Do you think the National Health Commission should organize a review of the preliminary planning, project brief, and design development files? It means to enhance the quality and refine the contents.

Tielin Wang: Once a project got approved, we would set up a plan to organize the relevant tasks. First, set the goal. Then, determine the schedule of construction. After that, we refine the whole project into sub-projects. Based on the refinement, contract plans can be drafted. We view these sub-projects as independent projects, such as those focusing on civil engineering or fire control. The whole project would be completed by combining these independent sub-projects. It's necessary to have a detailed budget on each sub-project.

Nowadays, once we got a project, if it were basically approved, we started working on project management: set the goals first, then make a schedule, and finally formulate a decomposition plan of the entire project. Moreover, there should be a reasonable cycle of project development. For instance, if the entire project needs to be done in two years, we should make decomposed schedules for the sub-projects. Once we had an overall schedule and a plan of decomposition, we would know how many sub-projects it should be divided into. Once these sub-projects were clarified, a contract plan should be drafted. This plan defines the sub-projects regarding infrastructure, civil engineering, or fire control as independent projects. Eventually, these sub-projects would be combined together as one project. The decomposition should clarify specific capitals for each project, that is to say how much investments each project would have, according to the index of costs. For instance, how much is the cost per square meter for a fire-control project or a construction project? In this way, the contract plan and the project schedule should be naturally consistent with each other.

A contract plan is based on the overall schedule and the decomposition plans. Once the contract plan was implemented, we would schedule the opening and completion of each sub-project. In the premise of scheduling the opening and completion of construction projects, we could also schedule those of infrastructure projects, as well as those of fire-control projects. These sub-project schedules are consistent with the time line for the entire project. As soon as the contract plan came out, a bidding plan should be made accordingly, which is in line with the time limits on construction. We typically conduct the project bidding three months in advance of a project opening and a half-year in advance for a relatively large project. Moreover, the duration between the completion of a project and the opening of a new one shouldn't be more than seven days, including the transfer time between two consistent projects.

If a bidding plan were not associated with the schedule of construction, the quality and schedule of the entire project would be infected. It's a cooperation. It is the management of tasks, such as which one should be done first, which one should be done in a specific month, and which one should be done at a certain stage. These should be in order. Once a biding plan were in place, the construction schedule, the contract plans, and the distribution of investments were in place. As soon as a bidding plan were submitted, the department of bidding put it on record. It clarifies how many biddings the project needs to recruit, 30 or 40, and how much the budget is. It is necessary to declare that biddings over the budget are invalid. It is the only way to control the costs. If a sub-project bidding is not in budget, the entire project will be out of budget. It means that capital flows, sub-projects, and biddings should be in plan. For project management, one thing of top importance is called "plan control", including the controls regarding project financing, arrangement, and biddings. Without a plan or a controllable plan, things would be out of control. In that case, the project would be in a passive situation, which requires additional investments. The schedule of construction should be reasonable and it is the prerequisite for bidding and construction. Project management consists of three main aspects: investment, schedule, and quality.

Lun Ge: Based on your experience accumulated in these years, do you think it is necessary to conduct an exclusive review on the diagram of system integration to ensure its quality?

Tielin Wang: The integration is truly important. Whether or not a set of construction drawings is completed may influence only one of the majors participating in the project. Contradictions among these majors make people worry since it would be difficult to overcome afterwards. So, we need to ensure the quality of drawings on each major and meanwhile take into account its relationships with other majors. Moreover, the standards for project design and construction, such as those regarding an intensive or basic interior decoration, should be clarified. I think a project review should be directed by its director and it is not a matter of a 'third' party. A review on construction documents is normative whereas a project review regarding its standards is a review by its director. When I say a director, I mean the one who is in charge of the project. Of multiple construction institutes, many aspects should be coordinated, such as the process of construction: which one should be constructed first and which one should be the following. These processes cannot be reversed. Secondly, the rules of construction should be followed, such as in-wall projects should go first and those outside of the wall follow. Similarly, there are rules of construction regarding building structures or civil engineering. If these rules cannot be followed, nothing would be achieved. Nowadays, hospital construction lacks rules and chaos are initiated by varied work methods or habits.

Regarding the bidding from a construction project, professional files and detailed documents are required and all of the contents should be clarified. For instance, how much would it cost for 250 window stills or stair treads? It's common to see that these items are missing in bidding documents. Everyone in the field of design and construction knows that elevation factors are of top importance in a project. If these factors could be clarified, more and more things would be unclear. People should develop their bidding documents by themselves. It is required and should be a classic norm to follow.

Lun Ge: In your opinion, how to evaluate the quality of a hospital project?

Tielin Wang: For a good project, we are in need of objective analyses and evaluations. Based a review of prior works, I think a technical analysis of project practicability could be the key, in addition to other important factors such as hospital management and maintenance. The technical analysis of practicability should be done during the process of construction.

Nowadays, there is a trend to compare Chinese hospitals with foreign hospitals and many hospitals want to be a 'first-class' or idol hospital. In my opinion, waste shouldn't be allowed either for interior decorations or regarding the standards of construction. *Suitability is of the most importance.* Technical analyses regarding project practicability should be conducted during construction.

For instance, the design idea of separating doctor flows from patient flows has been widely advocated. Where should we implement this idea? It would be reasonable to implement it in the design of an operating room but not an outpatient department.

If this idea were applied in an outpatient department, there would be many "dark rooms" and under-utilized spaces.

Practicability analysis regarding the standards for green buildings may solve the problem of "dark rooms". Regarding public spaces or offices larger than 2000 m^2, these standards set up clear requirements for natural lighting and ventilation. *According to the standards, 75% of the functional rooms in a hospital should have natural lighting. There are specific requirements for air-supply rates and energy consumption as well.* To my knowledge, we should firstly follow the basic requirements for construction and then take the peculiarities of hospital projects into account. Considering the peculiarities at the first place may fail to meet the basic requirements. Nowadays, a major problem of hospital construction in China is regarding the functional flows.

When compiling design requirements for functional flows, I summarized the flows at two level. Level one is regarding the relationships between the departments and level two is focused on the relationships inside a department. If the between-department flows were organized with each other in a reasonable way, the value of one plus one should be more than two. As a result, the distance that a patient needs to travel in the hospital can be reduced and staff's work efficiency can be promoted. Similarly, the functional spaces within each department should be carefully designed. The two-level functional flows can then be clarified. It wouldn't be a matter if all my words got deleted from this book but saving a single sentence saying "Technical processing of functional flows is the foundation of hospital design".

Lun Ge: Regarding the standards of technical processing, I would like to learn your thoughts on the 2008 edition of *National Construction Standards for General Hospitals.*

Tielin Wang: Regarding the standards, I raised an objection in the inspection meeting in Sanya, Hainan, many years ago. Why? Firstly, the standard floor area or land area per bed shouldn't be used as an executive index in practice. For instance, regarding the standard floor area per bed, it is 70 m^2 for a 300-bed hospital and 96 m^2 for a hospital with 800–1000 beds, which are relatively reasonable in current situations. Regarding the land area per bed, the current standards are 70 and 117 m^2 respectively. It is also reasonable. There is no need to further refine these standards. *A hospital should be an organic system. The deeper we investigate this system, the more we recognize the importance of utilizing integrated spaces. Every square meter should be used appropriately.* Large spaces are those supposed to be large and small spaces are those supposed to be small. Only when the general standards are appropriate, could we organize the areas of individual departments organically.

Secondly, smaller hospitals should have a relatively large proportion of floor or land areas per bed whereas larger hospitals should have a relatively small proportion, since generally there is room for large hospitals to make adjustment whereas that for small hospitals is very limited. If applying the same standards, the quality of small hospitals may get worse. As small hospitals are generally located in the areas with relatively low-level economic statuses, the budgets won't be much influenced if relatively high-level standards were applied.

Thirdly, I disagree about the classification of hospital spaces, such as the spaces for outpatients, medical technology, and inpatients, since this classification physically splits the organic system of a hospital. Based on it, designers need to group spaces by department, say, technology or inpatient. This doesn't contribute to the development of medical services as well as the overall architecture for health. A hospital should be viewed as an organic system. After planning the general system of a hospital, designers can then make specific design decisions on individual departments, which shouldn't be decided at the beginning. *How to define a good hospital? It should take a full advantage of every square meter as well as meeting the general standards and retaining its project characteristics.* The standards should be focused on the rationality of the whole system. Regarding detailed spaces, the standards of design should be regularized and flexibly accepted. For instance, a locker room shall be designed according to the size of its department and the number of staff members. *Hence, the standard size of floor area per person can be clarified.*

Lun Ge: I have a few more questions. Nowadays you are mainly engaged in construction management. How can you effectively connect management with design and minimize changes during the process of construction?

Tielin Wang: Typically, the main questions facing a hospital project are the following: firstly, project needs are unclear; secondly, the communication among project participants is not consistent; thirdly, the hospital fails to clarify their needs for technological spaces, especially regarding the equipment rooms, medical function rooms, and associated facilities. It takes a long time to coordinate a functional plan and a floor plan. The main reasons are associated with the interruptions received by the hospital during the time unifying different opinions. The hospital wouldn't like to take the responsibility and thus has no decision on it. As soon as the hospital made the decisions, the floor plans can then be settled down. At present, once the general outline of a hospital building has been decided, we would draft the layout of rooms at the first place and then send it to the design institute for professional design analysis and adjustments. It is important to ensure the schedule of design. From bid winning to design development, it generally needs half a year.

Lun Ge: How many hospital projects did you participate in? How do you feel about it?

Tielin Wang: There are about six or seven hospitals, including my first hospital project, the Heilongjiang Forestry Hospital, and the current one, Hainan Cancer Hospital (Fig. 3.22). I finally feel it is necessary to solve professional problems from a professional point of view.

Lun Ge: I have heard your talk regarding "solving professional problems from a professional perspective". It has become a classic quote.

Tielin Wang: *A project should be economic. Buildings don't need to be large but should be reasonable and appropriate.* It's necessary to simplify the decoration of building facades and keep the looking of a hospital plain and neat. Nowadays, people

Fig. 3.22 Aerial view of Hainan Cancer Hospital, Hainan Province

Fig. 3.23 Hainan Cancer Hospital, Hainan Province

intend to standardize hospitals, but it is very difficult. Do not have any wishful thinking that a hospital with standardized departments would be qualified. Actually, hospital projects should be individually specific. *At present, what we lack is experience, modes, and methods rather than money. We should clarify the differences between administrative and architectural issues.* Many designers are more likely to rely on experience other than rational thinking. Rationality means that we should identify the most essential things. Regarding our collaboration with oversea design institutes, there are something we could learn from them, but they shouldn't control the entire process of design (Fig. 3.23).

Lun Ge: What's your advice regarding the standards for hospital quality control?

Tielin Wang: The basic standards should be fully followed. *A hospital may have some 'personal characteristics' but not many.* Hospital design should be simple and neat. Secondly, the initial expectation for a hospital should be consistent with its long-term goals. *We could reserve spaces for sustainable development, but there are limitations.* As technology grows, the efficiency of hospital services is promoted. In Europe, attentions have been paid to economic growth. For instance, surgery centers are designed with double corridors, though single-corridor designs are also advocated. Such layout shortens the lengths of functional flows and saves a lot of floor areas. Some laboratory departments are equipped with automatic assembly lines and have achieved high levels of work efficiency. In fact, these don't necessarily require more floor areas. In addition, spaces should be reserved for teaching and scientific research (Fig. 3.24).

Fig. 3.24 Tielin Wang and Lun Ge

Lun Ge: Thanks for taking so much time and energy to answer our questions. We learned your morality, ideology, standing points, and tenacity. You are our idol. Thank you so much!

3.7 New Administration Patterns for Future Hospitals—An Exclusive Interview with Wei Zhang

Time: January 2015—reiterated in May 2023
Location: Guangdong Building, Beijing

Wei Zhang: Executive Vice President of West China Hospital of Sichuan University (Fig. 3.25).

Key sentences:

1. *According to Sun Tzu's Art of War, "Food and fodder should go ahead of troops and horses". Regarding hospital administration, we value not only the support of logistics to healthcare services, staff education, research, and administration, but also the positive influences of good logistics on hospital development, especially the facilitation and promotion of healthcare disciplines.*

Fig. 3.25 Wei Zhang

2. *Future hospital logistics is a broader concept of 'general logistics.' It requires the integration of all departments, refinement of duties, and sorting of procedures. The 'position' or role of an individual staff member would be more important than that of the department to which he or she were assigned.*
3. *Informatization is an important part of logistics management in which standardization, formulization, and informatization are interrelated with each other.*

Lun Ge: Nice to meet you, President Zhang. As the Vice President of West China Hospital of Sichuan University, you have been in charge of the logistics and construction for this hospital for many years. Your management is effective and has adopted advanced ideologies. Every time when I joined your presentation, I was impressed by your creative thoughts, specific ideology of administration, and advanced ideas. These aren't easy to acquire. We would like to learn the innovations you have made regarding hospital administration.

Wei Zhang: *I think the primary aspects include ideologies, administration modes, and informatization.*

In terms of ideology, first of all, we put an emphasis on hospital logistics. According to Sun Tzu's Art of War, "Food and fodder should go ahead of troops and horses". If we overlooked hospital logistics, things would be slowed down, including medical services, staff education, research, and administration. It is one of our main concepts on hospital development.

Secondly, we consider how to promote services, education, research, and administration via logistics, especially to drive and promote the development of disciplines. *The department of logistics is not a subsidiary but a facilitator of the development.* This is the ideology that our management team has had, since the early days of West China hospital. Our former president, Yingkang Shi, valued the importance of hospital logistics.

Fig. 3.26 Courtyard of West China Hospital Affiliated with Sichuan University (2023), Sichuang Province

Thirdly, *logistics planning should be in harmony with the ideology of hospital operation and administration.* For example, in order to provide convenience to outpatients, we should adjust the mode of services and facilitate interdisciplinary collaborations. As a result, a modular layout of clinics can be formed. So far, a lot of outpatient services still follow the "diagnosis-examination-diagnosis" pattern and have no updates for years. In my opinion, this is not only an issue of ideology but also a problem associated with hospitals' "hardware". Would an "examination-diagnosis" pattern be possible? If so, hospital design would be significantly changed (Fig. 3.26).

We initiated a mode of 'intensive' management to manage hospital logistics. It is different from the prior mode of 'centralized' management and doesn't simply put everything together in one place. In terms of management, we may use the 'intensive' mode. In terms of spatial arrangement, based on the number of patients and the estimation of service costs, we distribute medical equipment to varied places. It is quite efficient and economical. We should use the 'intensive' mode to manage clinical examinations or apply a hybrid mode combining the centralized and decentralized management. It's necessary for a hospital to have both large and small examination devices. To meet urgent requests, we may use convenient small devices; for a high volume of requests, we may use large devices to complete all the tasks. It is dialectical.

During the early development of our hospital, we applied the ideology of centralized management. After the buildings were built, we started developing new concepts. However, the new concepts were constrained by the layout of built environments. Consequently, there were a lot of transformation and reconfiguration since the buildings were built. For example, heavy devices for radiology examination were designed to be located in a large area on the first floor of the inpatient building 1#. They are now dispersed in the inpatient building 3#, 6#, and the Center of Physical Examination. If the hospital were crowded with patients due to an insufficient use of devices, we should immediately optimize the service management. The ideologies of 'centralized' management and distributed 'intensive' management should be constantly adjusted and enriched (Fig. 3.27).

Informatization is an important part of logistics management in which standardization, formulization, and informatization are interrelated with each other. As a matter of fact, certain tools of modern informatization can reduce the needs for human labor to a great extent, such as QR code scanning tools and service APPs. Even a decade ago, I suggested the pantry-men in our central kitchens to use palmtops accepting food orders from patients, and the logistical staff such as sanitary workers and security guards to use mobile devices connected directly with the security center. Informatization facilitates communication for problem solving. Moreover, problems involving with service procedures should be handled and optimized through informatization. This should be applied to every aspect of hospital development.

In addition, the structure of a hospital would be greatly changed upon the combination of standardization, formulization, and informatization. *Future hospital logistics*

Fig. 3.27 The main area of West China Hospital Affiliated with Sichuan University (2023), Sichuang Province

is a broader concept of general logistics. It requires an integration of all the departments, duty refinements, and procedure sorting. The 'position' or role of each staff member is more important than that of the department to which he or she has been assigned. By diminishing the traditional structure of management, we can set up a new system focusing on the key positions. It allows staff members to be multifunctional and reach their full potential. 'Higher efficiency and smaller team' will be the new mode of hospital management.

We have implemented this mode in our 1200-bed branch hospital. Based on its operating procedures and standards, we analyzed its management needs by using the tools of informatization. These tools assigned staff members to their positions and have presented high efficiency. For instance, the department of logistics only need 50 members in addition to the sanitary and catering staff.

Lun Ge: This is a revolutionary breakthrough. It is hard to imagine that the overall efficiency has been improved so much. Did you develop the standards and tools by yourselves? Can these be applied more broadly?

Wei Zhang: Our standardization and formulation were based the status of our environments, equipment, and infrastructure. These standards can be applied to other hospitals. The foundation of management remains the same when these are taken as the "standards" or "formulas". Specific modes of equipment may be different by hospital. Nevertheless, the key position of each equipment is the same. The ideology of management is the same.

The true reason why standardization can't be conducted in multiple hospitals is the missing of hospital cultures instead of standardization and formulation. If the culture of unity and concentration were developed, a hospital would be successful in promoting standardization and formulation. We are happy to see that hospital logistics has been quickly improved in China and believe hospital standardization and formulation will be developed soon.

In short, we should take logistics management as a scientific discipline and ponder all the steps and procedures. First of all, we ask the questions like "why would that be", "what it is", and "how to do it" and use scientific evidence rather than a sudden inspiration for analysis and verification. Moreover, we should be confident regarding the abilities of our logistical staff. Their levels of education may be relatively low but they have accumulated precious experiences in practice. We should stimulate their enthusiasm for work and encourage them to share experiences and turn practical experiences into ideologies and norms for management. This is crucial.

Lun Ge: It is of great significance. Logistics management is indeed quite intricate. You have gained a profound understanding in this regard. Do you any regrets in this field?

Wei Zhang: Actually, there are two things of regret in terms of our hospital management.

Firstly, we were too hesitant to try energy saving. For example, at the time of reviewing the proposal of Wenjiang hospital in 2007, we intended to use local geothermal resources to power the AC systems because of the very good geo-resources on site. However, the ideology of energy saving and green hospitals hadn't been generalized at that time and we finally dropped the idea of using geo-resources. It has been a pity that we missed the opportunity to apply the ideology of energy saving in practice.

Secondly, regarding hospital renovation, we should pay more attention to the step-by-step implementation of a developed plan. *Developing a long-term plan at the early stage of a project is especially important.* If there were no overall planning or step-by-step implementation, hospital development would be inhibited.

Besides, for hospital planning and architectural design, we are in lack of professionals and experts. These experts should be able to make systematic and integrative designs. *If we completely rely on architects to plan and design hospitals, it would be problematic since these designers may have very limited knowledge regarding medical services.*

Lun Ge: Thank you very much for your explanation. Next question is regarding your comments on our research of preliminary planning.

Wei Zhang: *Actually, preliminary planning is crucial to the development and construction of a hospital.* Nicely done, it does not only facilitate project biding, but also help to save time and money. After the hospital were built, there will be less needs for building renovation, leading to comfort and financial saving in a long run.

Therefore, I think that China needs to strengthen government policies in regard to the preliminary design and planning for hospital construction, including project management, project brief, and the process of medical services. Individual hospitals may not be capable to handle all of these issues. *Design professionals and management experts are needed to implement preliminary planning and design.*

Lun Ge: I couldn't agree with you more on preliminary planning. We introduce a tool titled 'SHAPE' for hospital projects' post occupancy evaluation (POE) last year. What do you think about it?

Wei Zhang: *The POE is necessary. The government should require it as a necessary procedure in hospital development. Evaluation needs to be done after a building has been occupied for two years. We can find deficiencies through a comprehensive post occupancy evaluation. It not only contributes to design refinement but also avoids a lot of sudden "inspirations" of hospital presidents.* So, POE is quite indispensable to the entire process of hospital development.

Lun Ge: Lastly, would you please discuss the specific influences of hospital development on architectural design?

Wei Zhang: Firstly, the design and planning of a hospital should be linked with its scientific development plan. It should be specifically scientific and meet the needs of target patient populations. For example, Wenjiang hospital is focused on rehabilitation services. It intends to provide patients 24-h access to rehabilitation devices and allows outpatients and inpatients to share the devices. To address these needs, the floor plans are designed to be large in size, say, about 15k square meters on one floor.

Secondly, the functions of a hospital building should go beyond healthcare services. It should be viewed as a general building for public use. *We should update our understanding of a hospital in many aspects.* For instance, we used to consider staircases as the exits for fire escape and locate them in the corners. However, the number of outpatients is generally large in hospitals and many wait for a long time to use elevators or escalators. The corner locations of staircases are inconvenient for them to use the stairs. From this perspective, a building for outpatients should be like a public building instead of a medical building. If staircases were openly located near to the escalators in the lobby, it would provide convenience to outpatients.

Thirdly, *we shouldn't overlook the needs of patient families.* The design of public areas for patient families should be based on the needs different by department. For instance, we should provide some waiting areas outside an operating room. In an intensive care unit, family members would be very anxious if they couldn't see their loved ones. Hospitals should also consider the needs and provide waiting areas for patient families.

Fourthly, in the process of construction, we should also consider how to set up supports for physicians, based on their needs to combine scientific research and clinical practice. An academic hospital has multiple functions such as providing medical services, teaching, and research. So, we should consider how to develop the service spaces that can undertake certain scientific research tasks, and how to develop the research spaces that also accept patients.

In addition, with the development of new technologies such as informatization, regenerative medicine, genomics, personalized treatment, and translational medicine, it's likely that we should make breakthrough changes in hospital design. For instance, with the development of informatization, more informationalized patient wards should be built for chronic care to facilitate follow-up services. From my point of view, we must update the modes of architectural design for hospitals. Besides aforementioned development, in accordance with the development of translational medicine, we may develop functional areas in hospital for integrated "outpatient services, laboratories, and patient wards". These would be an influence of the new healthcare services on hospital design.

Lun Ge: Thank you for sharing experience with us. Your distinctive perspectives and views on logistics management are meaningful, especially the model of 'intensive' management and its influence on hospital design and development. I'm deeply impressed (Fig. 3.28).

Fig. 3.28 Wei Zhang and Lun Ge

3.8 Steadfast Work and Focus on Data—An Exclusive Interview with Qixia Zhao

Time: January 2015—reiterated in May 2023
Location: Duty room at Beijing University Third Hospital

Qixia Zhao: Division Head of the Health Facility Management Department at Beijing University Third Hospital (Fig. 3.29).

Key sentences:

1. *In POE, hospital architects need to investigate how to design clear space and functional programs for medical services, follow the current standards, and thus promote hospital development.*
2. *Quantitative research regarding hospital construction needs to be supported by specific data, which in turn can provide evidence for POE. Using the quantified findings from data analysis can help to improve the quality of design.*
3. *Healthcare architects should pay more attention on the development of outpatient departments and medical technologies, and thus can objectively determine the appropriate proportions of floor areas for the major departments in a hospital.*

Lun Ge: President Zhao, you participated in the construction of Beijing University Third Hospital (Fig. 3.30). You also conducted a lot of exploring research in the field of hospital construction. Would you please share your experience with us?

Fig. 3.29 Qixia Zhao

Qixia Zhao: It is my pleasure to share experience. I think either hospital designers or managers would like to communicate with colleagues and share their experiences. I will be very satisfied if my experience could help others, either designers or managers, to avoid unnecessary detours in hospital design and construction.

In terms of hospital design, the growth of a medical architect is similar to that of a medical doctor since both need to accumulate experience in practice. What is a

Fig. 3.30 Beijing University Third Hospital, Beijing

high-quality hospital design? Although there are no standards to objectively evaluate the quality, patients' experience can answer the question.

Associated with promoted patient care and medical technologies, the bottleneck problem of hospital development in China is regarding inpatient departments. An inpatient department typically occupies a high percentage of floor areas in its hospital. Its size depends on the number of patient beds and the level of the hospital. Associated with recent improvements, the amount of time that patients spend in outpatient departments will be reduced dramatically.

From the perspective of hospital planning, I think it is too much for an outpatient department to take 15% of the overall floor areas in a general hospital, which is suggested by the *General Standards for Hospital Construction.* We may use first-hand data to introduce this perspective. During on our survey of the hospitals in primary cities, we saw many large and well-built outpatient towers were under-utilized. Meanwhile, inpatient departments were quite crowded in these hospitals and had no significant changes. I think hospitals should reserve enough spaces for inpatient departments to address the needs for development and properly downsize outpatient departments.

Moreover, I think an emergency department (ED) should take more than 3% of the overall areas, especially in a large comprehensive hospital. Why? Nowadays, Chinese typically view their medical needs as an emergency except the medical services traditionally provided in outpatient departments. So, ED is not just for patients who need emergency care but also takes the social responsibilities to provide elderly care and hospice care, etc. For example, an advanced cancer patient in need of hospice care may be accepted at an ED when approaching to the end of life. As a result, a lot of rooms in ED has actually been used to serve these kinds of patients. If ED has to be responsible for elderly care, hospice care, and rehabilitation, it must be a crowded area.

The third issue is regarding the distribution of hospital land. I think the current standards for land distribution, which require floor areas per bed, such as the *General Standards for Hospital Construction*, need to be improved. Although many cities are already short of land, we must make it clear that the current floor areas per patient bed are far less than those suggested in the standards for hospital construction. This phenomenon doesn't match the fast development in urban areas. From a perspective of research, according to *the Urban Land Use Classification and Standards*, the capital size of planned land for a hospital in urban areas should be defined at the 4th grade (105.1–120 m^2 per person). For instance, City of Shanghai adopts the standard of 120 m^2 per bed for hospital land planning. Based on our data analysis, I proposed 150 m^2 per bed in the article "*Key Issues in Comprehensive Hospital Planning and Construction*", which was presented in the 11th National Hospital Construction Conference. This can be used to update current standards and norms of hospital design and construction.

Lun Ge: What you have mentioned is very meaningful. We also conduct relevant research. Based on our research, we have proposed a POE (post occupancy evaluation) system titled SHAPE. Director Zhao, can you share your thoughts on POE?

Qixia Zhao: Data analysis for POE is very necessary. I am very glad to see your research and findings. Actually, the government has been advocating POE. According to the government document—*Management Approaches for Budget Control*, issued by the Ministry of Health, POE will be applied to ensure the quality of major projects, especially those with big government investments. In a prior version of this document which was implemented before 2013, POE after one-year occupancy was required for projects with an investment of more than hundred-million-yuan. However, due to practical issues, no such POE has been conducted so far. In my opinion, the government has seen the importance of POE but it is hard to technically quantify the evaluation.

The first challenge is the lack of specific evaluation criteria to guide the evaluation process. The second challenge is that if POE is based on the "*Management Approaches for Budget Control*," it mainly focuses on the evaluation components such as expected goals of the project, construction plans and engineering design, construction period, contract execution, and project investment efficiency (both social and economic benefits). The third challenge is the lack of standardized reference data for comparison of investment efficiency. Although POE is meaningful, the current evaluation methods have poor practicality. Nevertheless, this evaluation work is necessary and should be carried out.

Since the beginning of this century, a large number of hospital projects have been developed and are in need of POE. However, scientific measures are needed in order to evaluate project quality and frontier research is firstly required in order to develop the measures.

It is time to do these. The government has proposed POE from the perspective of project supervision and management. *In POE, hospital architects need to investigate how to maximize environmental functions for medical services, follow the current standards, and thus promote hospital development.* POE and building design are in an integrated process that requires our participation. Since a hospital building is a physical form, rational use of the hospital can maximize its ability to provide patient care. Quantitative research regarding hospital construction needs to be supported by specific data, which in turn can provide evidence for POE. Using the quantified findings from data analysis can help to improve the quality of design.

For example, in a hospital renovation and expansion project, the pharmacy for outpatients did not include any machines for medicine distribution. One year after occupancy, increased patient visits and limited storage spaces led to the implement of an automatic mechanical system for medicine distribution. It dramatically speeded up the medicine distribution without any building expansion. It also reduced the workload for pharmacy staff and saved time for patients.

It tells us that the add-on of just one medicine-distribution machine can significantly benefit a hospital. In prior, prints of laboratory report have been widely used in traditional hospitals and thus patients need to pick up their reports in the lab. Today, people can view their laboratory reports at home by using self-helped internet applications. The long queues waiting for laboratory-report pick up are gone. Supported by new technologies, both the physical spaces and labors can be saved for a hospital. *Healthcare architects should pay more attention on the development of outpatient departments and medical technologies, and thus can objectively determine the appropriate proportions of floor areas for the major departments in a hospital.* These require us to conduct POE from a new perspective of design and planning. By objectively investigating the actual situations of a hospital, we can develop convincing findings from POE.

We need to have a clear and persuasive process of data collection. It should respond to the specific functions of a hospital, regional discrepancies, patient habits, and system management modes. For example, regarding the standard number of patient beds in a general hospital, designers typically use the one-third of a 3-year average of daily outpatient visits. This method may look reasonable and be accepted by many government offices, including the Development and Reform Commission Department, Land Department, Planning Department, and health authorities. However, this method has been out of date since medical technologies keep updating patient care. Similar issues need to be analyzed through POE. *By collecting practical data on the quantities, sizes, investment, and operation of existing hospitals, a valid data base can be built and used as evidence to support hospital design.*

I had worked on the financial management of hospital construction for nearly 15 years. POE should be conducted from the perspectives of investment and operation. Based on current national policies on public investment, the collection and evaluation of data on the financial management in hospital projects is required. We should keep us updated in the field of hospital investment. However, we haven't done any actuarial analysis regarding hospital functions in terms of type and location. *Instead, we emphasize that hospitals shouldn't be over decorated and look like fancy hotels. What we have ignored is to promote the use of sustainable and durable materials for hospital construction.* We should evaluate these materials in the context of a building's life cycle and analyze their durability, levels of environmental protection, and impacts on medical care. We ask for objective estimates of the investment on durable materials for "green" hospitals. Regardless of the investment type such as social-based or government-based, the actuarial analysis and assessment on investment should be completed. Only in this way can maximize the benefits of investment.

Your hospital evaluation system SHAPE focuses on the critical issues of building design and technical processing for medical care. It is closely related to the medical functions of a hospital and the experiences of patients and physicians, which is very meaningful. I have a suggestion. Could you consider evaluating a hospital's land use in order to see whether or not its land allocation is reasonable? Whether or not the predicted volume of outpatients is accurate? Both the evaluation of land use and the assessment of medical services will influence the investment in this project. SHAPE has been tested in practice and is a matured theoretical system. Moreover, we need to help hospitals learn how to choose, analyze, and optimize data. It is our responsibility. Forty-four general hospitals that are in charged by the National Health Commission of

Fig. 3.31 Main Entrance of Beijing University Third Hospital, Beijing

China have shared their operational data for three years. Documentation and analysis of these data are critical and meaningful to our POE.

These data can also be used as research references. For instance, how large should an outpatient department be? We should investigate the data from the 44 hospitals, regarding the outpatient volumes, outpatient or inpatient department sizes, and inpatient admission rates. SHAPE can be used to facilitate the investigation, which may in turn help to refine SHAPE and make it more practical (Fig. 3.31).

Lun Ge: I agree and the availability of data is important. POE is critical and will become a hot topic. We notice that you have done a lot of research in this field. Your findings are substantial and supported by a large amount of data. How did you make it happen?

Qixia Zhao: Yes, data analysis is very important. For example, regarding the transportation inside a hospital, in order to update relevant design standards, we need data supports. How many parking slots on earth does a hospital need? To answer this question, we need quantitative data. There are a lot of factors influencing the needs for parking. Along with the rapid development of automobile industry in China, prior standards for parking design are no longer applicable. I was responsible to the revision of "Guidelines for Hospital Parking Systems—China Hospital Construction Guidelines" (2014). We collected data on the local standards for urban planning and construction management issued by 31 province-level capitals including Beijing and other 248 cities nationwide. Also, we sorted and analyzed the recent 4-year data of hospital parking in 159 cities. Based on statistical analysis of the first-hand data, we suggested a national average requirement for hospital parking design. This may help to solve parking problems in hospital.

Hospital transportation involves both dynamic and static situations. We need to consider both the dynamic traffic (e.g., road conditions) and the static traffic (e.g., the number of parking slots in a hospital and the parking demands from the surrounding communities). When we work on transportation planning for a hospital, a reasonable number of parking slots is a key limit that shouldn't be exceeded. Meanwhile, it requires an overall analysis of the traffics and a reasonable prediction of parking fees. Those are the examples of using and analyzing data in practice.

In general, as a medical architect and a project manager, one should abandon waywardness in order to conduct practical work. *We must apply our professional skills, consider the overall situation of a project, grasp both macro and micro issues, and then conduct a comprehensive communication to seek solutions to problems in hospital design.* We need to summarize hospital designs. That is the only way to improve and make progress. Telling people what you have been doing, what you have missed, and your regrets, is a way to prevent others from making similar mistakes. To be a responsible manager, you may summarize your work on hospital design and construction, and thus become a mentor in this field. In this way, we can better promote the quality of hospital design and construction (Fig. 3.32).

Lun Ge: Thanks for sharing your creative and thoughtful ideas!

Fig. 3.32 Qixia Zhao and Lun Ge

3.9 Respect the Basic Rules of Construction, Make Excellent Planning and Management—An Exclusive Interview with Yi Zheng

Time: January 2015—reiterated in May 2023
Location: Tea house in Yangqiao, Beijing

Yi Zheng: Former Assistant-President of Beijing Ditan Hospital affiliated with Capital Medical University (Fig. 3.33).

Key sentences:

1. *The first lesson we have learned is that we must respect design and the basic rules of construction and leave a plenty of time for preliminary planning and argumentation.*
2. *It is very necessary to spend more time in preliminary planning, in order to make the planning detailed, the argumentation sufficient, and the proposed functions complete.*
3. *Renovating a hospital was even harder than building a new one. In order to completely solve the problems, we need a complete reform of our hospital. We should have foresight in project planning, project scale, investment, and many other aspects in order to build a truly sustainable hospital.*

Lun Ge: Hello, President Zheng! You have participated the entire construction of Beijing Ditan Hospital (Fig. 3.34). Would you like to share your experience with us? Any feedbacks since its operation?

Yi Zheng: Generally speaking, Ditan Hospital is a relatively modern and advanced hospital. It is also a successful case of whole-hospital relocation. It was put into operation soon after it was built. It guaranteed the emergency medical services for the 2008 Olympic Games and later responded to the challenge of A/H1N1 flows. In terms of its advantages, it has played an important role to protect our capital's security and public health, and also made good achievements in the prevention and control of infectious diseases. All of these are worthy of recognition.

Fig. 3.33 Yi Zheng

Fig. 3.34 Front view of Beijing Ditan Hospital, Beijing

In addition to these advantages, we must face the shortcomings as well.

Conducting discussion on the lessons we learned may be more beneficial than talking about the achievements. For example, the building volume is too large and there are too many dark rooms. The inpatient building has three rows of rooms and four corridors. In the middle row, there are physicians' offices, nurse stations, and medical rooms. These room are dark rooms without natural ventilation or lighting. They rely on mechanical ventilation and artificial lighting, which increase not only energy consumption but also operating costs. Even if these devices are well adjusted to perform the best, the environment always makes a sense of oppression.

Working in such an environment for a long time will affect not only the staff members' physiological and psychological health but also their work efficiency. Hence, we must try our best to avoid designing this type of environments in future hospitals. There is no need to over pursue a big building volume. One of the reasons for these problems is that we conduct the design in a rush and there is no sufficient time for argumentation and comparison, especially in the early design period.

Therefore, the first lesson we have learned is that we must respect design and the basic rules of construction and leave a plenty of time for preliminary planning and argumentation. However, in this project, the duration of time for design was only 6 months. We had to use the preliminary design to invite bidding for construction. It is against the basic rules of construction to have such a short time between design and construction. However, the situation was very special and the project was required to be put into operation in 2008. The schedule was extremely tight. I have repeatedly emphasized the importance of respect to design.

Meanwhile, we should respect the procedures of design and the basic rules of construction. Otherwise, the hidden dangers and malpractices would have long-term negative effects, which we have deeply experienced in the daily operation of this hospital.

Then, let us discuss our experience regarding project planning and management. In 2002, Ditan Hospital was included in the list of institutes under consideration for shutting down and reorganization. This hospital would be withdrawn from the list of government-affiliated institutions and thus there would be no Ditan Hospital any more. The entire hospital was extremely down and everybody was in a panic.

Due to the outbreak of SARS in 2003, the government decided to not only cancel the shutting down of Ditan Hospital but also strengthen it to be a specialty hospital focusing on infectious disease care. In April 2004, the Beijing Municipal Government built a team to relocate and rebuild Ditan Hospital. With an investment of 420 million yuan, it was planned to be a hospital with 600 inpatient beds and 800 daily outpatient visits. Only a few years after occupancy, the number of inpatient beds in this hospital has increased to 758 (26% increase) and the number of daily outpatient or emergency visits has peaked up to 2248 (181% increase on average).

How much is the pressure if using the designed capability of 800 visits to meet the actual need of 2000+ visits? How to ensure the quality of patient environments and the work environments for staff? The hospital must expand itself to cope with this situation.

However, we were short of a series of equipment for expansion, such as electrical transformers, sewage treatments, and oxygen supplies etc. Renovating a hospital was even harder than building a new one. In order to completely solve the problems, we need a complete reform of our hospital. We should have foresight in project planning, project scale, investment, and many other aspects in order to build a truly sustainable hospital.

Lun Ge: President Zheng, can you please talk about your thoughts regarding the agent construction systems?

Yi Zheng: The prevalence and constant phenomenon of excesses in hospital projects, such as the un-necessary increases in project scale and exceeded investments, made the management side worried. Construction companies were thought to be careless about government investments and thus a third party good at construction should be identified to do the job instead.

In 2004, Beijing Development and Reform Commission issued the *Management Approach of Agent Construction Systems for Beijing Municipal Government Invested Construction Project* (Trial version, issued by Beijing Development and Reform Commission, No. 298 document). Ditan Hospital became one of 'the first persons who eat crabs.'

Agent construction systems are new to everyone. The government hoped to have someone who knows construction to do the work for hospitals, but the complexity and particularity of hospital construction hadn't been fully noticed. *Understanding construction does not mean understanding hospital construction.*

In addition, at the beginning of this project, a series of policies and rules of the agent construction systems were not optimized. The construction agency was in charge of everything. However, other government agencies didn't recognize the power of the construction agency and still asked for approvals from the construction practitioners or companies.

As a result, here came a weird situation: the construction agency made decisions but the construction practitioners need to take all the responsibilities. This caused certain problems regarding the entire agent construction process.

The initial willing of agent construction was good but it gradually got out of the way. For the sake of financial benefits, many agent construction companies ignore the interest of society and the needs of hospitals. As a result, they have caused negative influences on hospital design and construction.

For example, operating rooms (ORs) must be professional and with high technologies and functional intensiveness. It is expensive. Agent construction companies emphasizes on money saving. They broke down the integrated systematic rules of OR design and assigned the design tasks to different contractors. They made separated tasks regarding the equipment, civil engineering, pipelines, and central control systems, etc. For instance, a project of ventilation pipes was assigned to three contractors. It was hard to identify who should be responsible for the pipes if there were a leakage or insufficient air supply problems.

Similar things happened regarding the control panels on OR room walls. In order to save money, these panels were taken out of the central control system and purchased separately by the agent company. As a result, these panels encountered a lot of trouble in the comprehensive debugging. For many times, hospitals tried to communicate with the agents and emphasized on the importance of integrity, technology, and equipment linkage in ORs. However, it did not work. Several operation indicators in the ORs failed to reach the acceptance criteria and eventually it was the hospital who had to take the consequence.

The original intention of developing the agent construction systems was good, but the systems were incomplete in some parts such as the missing of a legal status of construction agencies, the reorganization of the agencies from multiple government departments, and so on. Especially, in the context of the No. 298 document, if there were financial savings from a hospital construction project, more than 30% of the savings would be rewarded to the construction agency. Maximizing profits would be the highest purpose of the construction agency. Coupled with such a policy orientation, there would be more magnified profit-seeking of the agencies and the consequences can be imagined.

With continuous improvements in recent years, the agent construction systems have been refined. It is very important to engage people who is professional at hospital construction to manage hospital construction. *Let professionals do professional work.*

We should coordinate the relationships between the agencies and the construction practitioners, make a better control of government investments, build high-quality hospitals for the public, reduce hospitals' burden, and go back to the original intention of developing the agent construction systems. The systems will go better and more stable.

Lun Ge: It seems that the government still needs to refine the agent construction systems and there is a long way to go. Since Ditan hospital was put into use, regarding the initial project goals, which have been achieved and which haven't? What do you think about the post occupancy evaluation (POE)?

Yi Zheng: In current management systems, preliminary argumentation and POE are insufficient. The importance of pre-design planning and argumentation has been discussed a lot. *It is very necessary to spend more time in preliminary planning, in order to make the planning detailed, the argumentation sufficient, and the proposed functions complete.* Then, the construction process will be smoother with less demolition and reform, leading to a better control of the investment. This has received more and more consensus and attention. Starting a hospital construction project with a design rushed out in 6 months and launching the bidding with preliminary design documents are not preferred. I really hope these won't happen again. This is a serious violation of the basic rules of construction. Without the serious, responsible, and countless efforts of many experienced colleagues and our team members, more regret of Ditan Hospital would have been happened.

Regarding the post occupancy evaluation, it is even worse. When one project were finally finished and all the money were spent, have we reached the original goals? Do we have any social benefits or economic benefits? Is the hospital a good one? Are the physicians satisfied? How about the patients? There are no answers. Your project about POE is very timely. *The proposed five aspects of SHAPE (security, humanization, architectural function and space, physical environment, and efficiency) are excellent. I think the evaluation should be focused on not only the design drawings or design itself, but also the entire construction project.* It is very meaningful and groundbreaking.

I suggest to add an evaluation on the effectiveness of investment. For example, the government has spent so much money on a hospital project, do we achieve the goals? What are the benefits from it? Are the investments and the outputs proportional?

Therefore, the POE should include not only the five aspects of SHAPE but also the economic benefits of the entire system. Then, the evaluation framework would be more complete. If confirmed by law as a must-do procedure, POE will bring a great promotion to hospital construction (Fig. 3.35).

POE can also help the government's functional departments to analyze the reasons why the number of daily visits to the outpatient and emergency department was about 800 at the time receiving the project approval, but has increased to more than 2000 in the built project.

Lun Ge: Your suggestions make practical sense on the POE. We will continue to work on it. We would also like to hear your opinions regarding the dual-corridor design in hospitals.

Yi Zheng: Nowadays, there is a trend to set up dual-corridor systems in hospitals. The intention is to reduce unnecessary contacts between physicians and patients in order to protect physicians from possible conflicts. It is reasonable to set up dual

Fig. 3.35 Courtyard of Beijing Ditan Hospital, Beijing

corridors under certain conditions, such as in infectious disease hospitals. However, in general hospitals, dual corridors may lead to a waste of resources.

It is really good to have a consultant room with natural ventilation and lighting. Practice has proved that natural ventilation is a simple but effective and low-cost way to reduce nosocomial infections. With a dual-corridor system, chances for the interior to get direct natural light and ventilation are reduced. Meanwhile, it leads to increased needs for ancillary mechanical equipment, more energy consumption, maintenance, and costs. It also increases the size of areas in demand.

The proposed protection for physicians is soft. No matter how many corridors there are, physicians and patients meet in consultant rooms. *The best way to solve the conflicts between physicians and patients is by law.* A dual-corridor design may reduce the access to nature and make physicians' work environments even worse. *These concerns should be considered based on humanization and functionality, which could be a good research question.*

Lun Ge: President Zheng, thank you very much for the sharing! We appreciate (Fig. 3.36).

Fig. 3.36 Yi Zheng and Lun Ge

Chapter 4
Interviews with Hospital Architects

4.1 Exploring in a Complicated Situation—An Exclusive Interview with Guoliang Chen

Time: January 2014—reiterated in May 2023
Location: The office of Shanghai Architectural Design & Research Institute Co., Ltd., Shanghai

Guoliang Chen: Chief architect of Shanghai Architectural Design & Research Institute Co., Ltd.

Representative work: the Xiamen Branch of Zhongshan Hospital affiliated with Fudan University, Fujian Province (Fig. 4.1).

Key sentences:

1. *I think the functionality of a hospital street should be preferred than its symbolic value. It serves as a hub connecting the individual functional areas for medical services. It's part of the transportation system, which may have diversified forms.*
2. *It would be quite problematic if hospital design and construction were not carried out either from the inside to the outside or by integrating the interior and exterior design.*
3. *If we invited oversea designers for hospital projects, they should remain in the team during the schematic design phase, complete the primary and secondary level of design on functional flows. What we need is their experience designing the hospital as a whole project rather than a mere shell.*

Lun Ge: In your opinion, what are the problems in today's hospital design and construction? We think the stage of preliminary planning is relatively weak. A lot of problems occurred at the later stages are associated with inadequate preliminary preparation.

Guoliang Chen: I think preliminary preparation is not only necessary but also crucial. As members of the Medical Building Expert Advisory Committee of the National

L. Ge and Z. Wang, *Thoughts on Hospital Design and Construction in China*,
https://doi.org/10.1007/978-981-97-0818-5_4

Fig. 4.1 Guoliang Chen

Health and Family Planning Commission, if reviewing a hospital's proposal for construction, we are required to firstly understand the hospital's development plan, including its positioning, target patient populations, scale, scientific disciplines, etc.

These make the foundation of technical documents, on which we review the overall planning of the hospital. Only in that way could we know how to implement the determined positioning, scale, classification, scientific disciplines in the overall planning. This is needs-oriented design. The last thing in the review is to check out the individual construction projects. None of these steps is dispensable.

How to judge the quality of a hospital design? Judgements would be unscientific if we did not know the needs of the hospital. It is unscientific to judge a hospital based on whether or not its appearance is attractive or the indoor environment is modern enough. The key to succeed in hospital design lies in whether the hospital can present specific and clear needs and whether or not the design could meet the needs.

One key aspect of post occupancy evaluation (POE) is to investigate whether or not the aims set up in early design have been achieved in the later stages. This is of great necessity. I think many hospitals are deficient in professionality, normativity, and many other aspects.

Lun Ge: This is one of the significant purposes of POE.

Guoliang Chen: Recently, we met several investors and design teams from foreign countries. They would spend eight or nine months to research the positioning of a hospital. Based on the research, the hospital's service disciplines and the scale and development of each discipline would be set up. The planning of medical equipment configuration is then carried out. Eventually, they calculate the proportion that each functional area requires and the overall size of floor areas. Hence, they will be able to design a functional layout compatible with the hospital's positioning. I think we should learn these methods.

Lun Ge: It's scientific and promising. Let's lead everyone towards this direction. Here is a second question. Is it necessary to review the brief of a design?

Guoliang Chen: I think it is quite necessary. However, to implement the review, it is critical to figure out who the reviewers would be. I am afraid that some reviewers themselves may not understand hospital design.

Lun Ge: Moreover, before starting the work for design development, do you think it is necessary to review the proposed technical process of medical functional flows and relevant equipment systems?

Guoliang Chen: I don't know the design regulations applied in Beijing. Those in Shanghai can be used as a reference. The regulations applied in Shanghai requires a project with certain levels of technical challenge to be reviewed before the stage of design development, by experts organized by the Science and Technology Committee and the Construction Committee. Then, the project shall be refined according to experts' comments. Only after that, applications can be submitted to the government for an approval of design development. I think this is very similar to your idea. As the design and construction of a hospital involves in complicated technology, it is very necessary to conduct a technical review before issuing an approval of design development, in which the technical process of medical functional flows and relevant equipment systems may also be included.

Lun Ge: Exactly. We should strengthen the quality control of architectural technology applied in hospitals.

Guoliang Chen: I suggest to conduct this type of review for selected hospitals as the first step. There is no need to cover all hospitals. We may start reviewing the hospitals in a super-large scale or those requiring particularly intricate technology. Upon positive feedbacks, possibly, more hospitals would join.

Lun Ge: It's wonderful. This will lead hospital development into a virtuous cycle, which is also our original intention.

Guoliang Chen: Generally, our hospital projects are large in scale and thus each of them needs to be reviewed by the Construction Committee and the Committee of Science and Technology. I quite agree with you. The panorama described in your research includes several design stages and detailed contents by stage. At the beginning, it's difficult for us to refine design in detail. The positioning of a hospital is of top importance, such as a hospital at the national, regional or local level. According to the positioning, we can then estimate the number of general populations and patient populations, and the spectrum of diseases for hospital design. Then, the disciplines of proposed medical services can be specified. I think these are the core principles of hospital design. As the process of design goes forward, what to do at each design stage and the contents and standards for individual subsystems will be gradually clarified and specified.

Lun Ge: In this panorama, we divide the process of design into four stages. Colors representing these stages in the research graph turn darker as they get advanced in the design process. During my participation in the reviews, I frequently felt that some stage-one issues were brought up in the final stage. As a result, these issues couldn't be clarified and solved but took a lot of time for discussion in the reviews.

Guoliang Chen: Hence, I think there two steps to establish a framework of hospital design: one step is to structure a medical service system and the other is to specify the problems that need to be solved at each design stage. As you mentioned, it would be very helpful to develop some design templates.

Lun Ge: We have earned rich experience in this field. There are no normalized, scientific, and rational standards or guidelines for designing a national-level medical infrastructure. We have some experience but it is hard to provide specific guidance to individual projects. If the design team and the investor were experienced and responsible, the quality of a hospital project would be guaranteed. Otherwise, it would depend on luck.

In the stage of construction document development, is it necessary to conduct a review of the medical technology design? It is regarding the purification systems, functional layouts, light current systems, logistics, and transmission systems, etc.

Guoliang Chen: At present, at the stage of design development, we have three aspects of design: (1) air purification systems, light current systems and logistics, (2) terminal-point locations, and (3) construction standards. For instance, high levels of cleanness required for operating rooms, the form of air conditioners and the overall capacity will be determined at this stage. Then, we conduct technical discussions with the hospital to estimate the cost of these items and include it in budget. By doing these, with the completed documents of design development, the hospital could commence biddings for technical projects. It is crucial to have a technical review in this regard, but I don't think it's necessary to do an exclusive review of technology. We may integrate it into the review of design development.

It's the current condition in China: Compared with other countries, China has fewer professional consultants who can collaborate with architects in hospital design. Therefore, it's often the case that professional contractors (most of them are qualified for specialized technical design) are responsible to the technology design of a hospital. Architects are responsible to review these contractors' design work, majorly focusing on the design standards set by building and technology system, to ensure that individual technical systems are integrated with each other. There are no accrediting agencies between architects and professional contractors.

Lun Ge: **At** this stage, you conduct an internal review to ensure the continuity of design.

Guoliang Chen: It's true. We review and sign on the drawings submitted by the contractors but are only responsible to the design standards and the building interfaces of medical systems. Just like the design of a glass curtain wall, there is a structural interface connecting the curtain wall to the main structure, which is reviewed by us to ensure its quality. Meanwhile, the specialist contractor is responsible to the curtain wall system. Same thing applies to the air purification systems. If the level of cleanness didn't meet the purification standards in practice, it must be related to the inappropriate standards set by the hospital designer. If the hospital lacked resources for heating or cooling, it would also be the designer's responsibility. However, if the purification conditioning system itself had problems, it should be solved by the contractor.

Lun Ge: I wonder if this is part of your routine management.

Guoliang Chen: As far as I know, large domestic design institutes specializing in hospital design prefer to manage their projects in this way. However, quite a few of them failed to do so.

Lun Ge: So, the stage of design development is really of great significance.

Guoliang Chen: No matter we as the general contractor outsource the project to subcontractors, the construction standards and system configurations should be determined during the project review of functional flows and systems. Take an operating room for example. What kind of surgeries are to be performed there? How much is the initial investment? Are we capable of undertaking the cost of daily operation? All of these require intensive research and verification. Only on the basis of that could we identify the reasonable level of cleanness for this design. Another example is the design of a light current system. We list all kinds of light current systems commonly used in hospitals before entering the stage of design development. There could be over 20 systems. Then, we inform the hospital that certain systems would be necessary, such as the security system and the firefighting system, whereas the rest are optional. Whether or not taking these optional systems into design consideration may depend on the hospital's development plan and operation mode. While, some systems may be reserved for future development. The stage of design development is carried out in accordance with the established principles of design.

Lun Ge: I have another question. Do you think it is necessary to conduct the interior design at the stage of design development? If not, interior designers would present their designs after winning the bid for construction, which may lead to a reengineering of functional flows.

Guoliang Chen: It's appropriate for interior design and architectural design to run in parallel or to start the interior design at the stage of design development. We have learned by working with overseas design companies that they have interior designers involved in early design. In many joint projects, design institutes from foreign countries may leave the construction drawings for domestic design institutes to complete, but often complete the work of interior design at the stages of schematic design and design development.

Besides, hospital interior design has its uniqueness since high-standard interior decoration is needed for only 20% or less of the built areas in a hospital. Primarily, these areas require few medical functionalities, such as the outpatient department, inpatient hall, reception center, VIP lounge, and multipurpose rooms, etc.

There is still a large part or 50–60% of the hospital areas has strong needs for medical functionality, which should be completed either by architects or the cooperation of architects with interior designers in which the former takes charge and the latter supports.

As to us, we usually hire a professional interior design institute to work with us at the later stage of design development. The interior designer proposes design styles and tones for discussion. After receiving a confirmation from the owner, the interior design institute takes charge of the design of about 20% of the total areas, from the schematic design to design development and then construction documents. The rest areas (about 50% or so) are in the charge of the architects. They refer to the selected interior design style in the development of construction documents to ensure the continuity and consistency of design style and to meet the requirements for medical services. *If the work of interior design were started late, any changes of the interior design would result in endless rework of architectural design and reconstruction.* It's certainly the best to take interior design as part of a unified design management system, which has been implemented in many of our projects.

Lun Ge: You are the general contractor and the contracted interior design institute would be a subcontractor.

Guoliang Chen: Regarding the question which interior design institute should be selected, a joint decision is made by us and the client. We are the party to sign the contract of interior design, which is under the unified management and coordination of architects.

Lun Ge: Your experience is worthy of popularization. In most construction projects, it's still the case that interior design is launched after the completion of building structure, leading to numerous design problems.

Guoliang Chen: The vertical and horizontal relations of design development to other design stages are vital. It's of great necessity to strengthen the review of design documents at this stage. Also, it's a very important procedure to ensure the quality of hospital design and construction (Fig. 4.2).

Lun Ge: Which design stage is more rational if conducting research on project feasibility? I interviewed President Tielin Wang. He thought it would be appropriate to conduct the research after the stage of design development.

Guoliang Chen: His thoughts make sense. Nevertheless, in terms of design continuity and maneuverability, I think it would be better and more appropriate to conduct the feasibility research when two thirds or at least half of the design development were completed.

Currently, we suggest clients to work on their project proposals and feasibility reports separately. To develop a project proposal, a hospital may entrust a professional consulting firm to provide a general site plan with major indicators, which is quite simple. Once the proposal got approved, the hospital may organize a bidding and choose a design institute for this project.

Fig. 4.2 Xiamen Branch, Zhongshan Hospital Affiliated with Fudan University, Fujian Province

However, a report on project feasibility requires detailed information and should meet the specific and explicit requirements set by the *National Development and Reform Commission (NDRC)*. Research on project feasibility should be completed on the base of an approved project proposal and when the design development has reached to a certain level of depth. During the proposal review, managers would make specific requests of which some may affect not only the design but also the investment.

At the stage of design development, the first one third of time is spent on quantifying the basic systems in design, such as the structure, electromechanical, etc. Only based on that could we work out the accurate amount of project investments and have a reasonable analysis of the cost. This is the most important part of a feasibility report.

If the research on project feasibility and relevant reviews of the report were performed after the completion of design-development documents, the development of construction documents would be delayed, leading to significant impact on the overall schedule.

Lun Ge: Your approach is really advanced and scientific.

Guoliang Chen: If a feasibility report were done too early, findings from the research would be unspecified. It would be hard to tell its scientificity and feasibility as well as the rationality of the budget. It is of the most importance in review.

Lun Ge: In 2010, we initiated an important research project funded by the Beijing.

Natural Science Foundation. Based on this research, we put forward a post occupancy evaluation system titled "SHAPE" and would like to hear your thoughts on it.

Guoliang Chen: *SHAPE provides an excellent idea to evaluate a hospital project in five domains, which have covered all the fundamentals.* It is very nice but I wonder if the workload would be heavy. You have included every item concerning hospital design. The workload for evaluation would be heavy.

Lun Ge: The workload is heavy indeed, but it's of great significance. We have formulated a theoretical framework, the evaluation criteria, contents, and methods of evaluation. During the design and construction of a new hospital, there are neither quality control mechanisms nor evaluation standards. Most hospitals aimed at constructing landmark buildings ranking at the first level Asia–wide or worldwide. Whereas, it's hard to decide whether these buildings are qualified. How could they meet the standards of being a world-leading or Asia-leading?

Guoliang Chen: What you are doing is truly meaningful. There is certainly nothing wrong to aim at building a world-leading or Asia-leading hospital at the beginning of a project. However, the understanding of a 'world-leading' and 'Asia-leading' hospital may be different. Since the policies of reform and opening, China's comprehensive strength has been rapidly improved. We have made remarkable achievements in promoting both the scale and the quality of hospital construction and have accumulated a lot of experience. However, there are still a considerable number of low-level design and low-quality hospital projects. It's time to setup the standards for "a genuinely good hospital."

Generally, we won't attend any bidding for a hospital project, if the hospital had no impartial and scientific evaluation standards. If a design proposal won for its unique, weird, exaggerated outlooking, a certain idea, or a slogan that impressed the hospital president, young architects would get lost and confused about the origin principles of hospital architecture.

Lun Ge: At present, the only criterion of hospital projects in China is the 2008 version of *Construction Standard for General Hospitals*. Peers in this field have divergent opinions on it. Please share your thoughts.

Guoliang Chen: In fact, our institute participated in the editing of this 2008 version of standards. It provided a basis for regulating the design, investment, and construction of hospitals in China. It is of great value in terms of providing a basis for national and local competent departments to issue project approvals.

However, with the continuous development of hospital construction, certain parts of the standards need further discussion and improvement. As far as I can see, there are two major issues.

One is regarding the system of "standards." So far, we estimate hospital scale by using a single indicator: the number of inpatient beds. Based on this indicator, we acquire the rough size of a hospital's floor areas according to the standardized size of areas per bed. By adding a certain size of affiliated areas, the total size of a proposed hospital can be made up. In this way, a hospital's size depends merely on the number of inpatient beds.

However, the volume of outpatients, the type and quantity of daytime medical services (such as day surgery, radiotherapy, and chemotherapy, etc.) vary by region or hospital technology level. Furthermore, the positioning of a hospital differs a lot from each other. For instance, is your hospital focused on providing primary medical services for a certain region, curing miscellaneous diseases, or providing specific care on oncotherapy? Associated with different positioning, the requirements for floor areas are different by hospital. However, during the review for a project approval, the competent departments often calculate the size of a hospital merely based on the number of beds in accordance with the standards.

They commonly divide the floor areas of a hospital into 7 functional groups and one group of additional items. This is unreasonable since these groups are individual entities but a hospital is one system. I suggest to introduce a multidimensional index or parameters in the review, together with the index of bed numbers, to determine the scale of a hospital in a more scientific and reasonable way.

The other issue is related to specific provisions. The most controversial one is regarding the standards of land use. The code titled *Construction Standards for General Hospitals* has stipulated explicitly that the standardized size of a general hospital is about 109–117 m^2 per patient-bed. However, it typically needs to be 110–130 m^2 per bed without counting in the areas of underground parking and so on. Some areas can be located underground so that the plot ratio turns out to be merely 0.8 or so. However, the plot ratios required by the standards of hospital planning range from 1.5 to 1.8. Why is this criterion of land use so high for hospitals? Someone would say it is for future development. I somewhat disagree with them.

If a hospital's capacity were proposed to be 1000 beds, the hospital should then acquire or reserve the land in accordance with the land-use index of 1000 beds. If the first phase of construction planned to build 500 beds, it would only be the plan of construction for the first phase. Otherwise, it would be hard to meet the requirement of land-use index for future construction. So, I suggest using the index of plot ratio to control hospital land use.

Lun Ge: I agree with you. I think there are no systematic standards. This is exactly the reason why we categorize the floor areas of a hospital into seven domains by function and spatial use.

Guoliang Chen: Your classification is relatively comprehensive and scientific. Moreover, each of the domains or functional systems corresponds to a group of rooms.

Lun Ge: During our interview of President Tielin Wang, he suggested using the overall size to control hospital size. According to him, it is only necessary to determinate the total size of a hospital whereas the specific proportions of individual departments would depend on the hospital's positioning.

Guoliang Chen: I totally agree. For instance, the concept of modern emergency care is different from the traditional manner of emergency services, which has been

Fig. 4.3 Entrance of Xiamen Branch, Zhongshan Hospital Affiliated with Fudan University, Fujian Province

applied in many hospitals to build their emergency centers. How could it be appropriate if there were only 3% of the total area distributed for emergency care? The ratio should be doubled at least (Fig. 4.3).

Lun Ge: Do you agree on the suggestion of separating the parking area as an independent design item?

Guoliang Chen: In fact, regarding hospital transportation, relevant technical requirements and indicators are clearly stipulated in the *Regulations on Urban Planning and Specified Planning with Controllability*. Whether or not there are social resources for hospital parking depends on local economical levels. The situations in developed areas are clearly different from that in less-developed areas. In addition, it's also unreasonable to plan parking spots at the same ratio (1:1 or 1:1.5) for hospitals with different capacities (500 or 1000–2000 beds).

Lun Ge: I have another question regarding the interlayer space between hospital floors. According to the standards, the height of an interlayer space should be 2.2 m otherwise workers couldn't climb in for maintenance. If in case the height went beyond 2.2 m, which functional group would the additional areas belong to?

Guoliang Chen: In this respect, I think the voice of the hospital authorities is a little bit weaker than that of the planning authorities. Due to the particularity of medical buildings, equipment interlayers are very common. The height of 2.2 m is very hard for equipment and pipeline installation, as well as daily maintenance and overhaul. Therefore, we recommend that the areas of a technical interlayer used only for arranging equipment should not be counted in the total floor areas, as long as its height is less than the height of a standard floor. If it must be counted, it should be included in the areas of a logistics support department. Of course, the allowed areas of logistics support should be increased appropriately.

Lun Ge: Regarding the size of a hospital street, the standards didn't mention. However, the street areas are getting bigger and bigger. Which system should these areas belong to?

Guoliang Chen: It's true. Hospital street areas get much larger than they used to be. Many hospitals have viewed the streets as a symbol of modernization and high-grade. *However, I think the functionality of a hospital street should be preferred than its symbolic value. It serves as a hub connecting the individual functional areas for medical services. It's part of the transportation system, which may have diversified forms.* It is suggested to list the street areas as part of the hospital public transportation systems so that we can make a more scientific review and rational evaluation of it.

Lun Ge: As hospitals have certain attributes in common, we reclassified the qualitative and quantitative criteria into two groups: fixed criteria and open criteria. Do you think it is acceptable?

Guoliang Chen: I completely agree with you. As to the reason, if there were no specific limitations, the standards would be too flexible and nonbinding. On the contrary, were they completely fixed, the standards would be ill-adapted to the peculiarities of each hospital. The way you deal with it is somewhat like the existing *Construction Standard for General Hospitals*. In the standards, there are basic floor areas and sub-items that can be listed and calculated separately. In this way, the standards applied at the stage of preliminary planning and in the post occupancy evaluation are consistent.

Lun Ge: Meanwhile, we are firmly against the concept classifying hospital functional departments into isolated buildings. This concept suggests to build one building for each of the functional departments. In that way, spatial arrangements would be disconnected from user needs.

Guoliang Chen: I am in full agreement with you.

With the development of medical technology and constant changes in medical disciplines, the departments of outpatients, inpatients, and medical technology are more and more closely integrated with each other. The functionality of emergency care also grows much more prominent with well-equipped supporting systems. Moreover, many hospitals built independent emergency centers. Accepting the introduction of modern logistics systems, some hospitals arranged their auxiliary departments (such as the central laboratories) in remote locations. These departments are scattered in hospitals. Examination results issued by the laboratories can be transmitted through the logistics systems and presented on physician's computer screens. This facilitates patient visits and makes it possible for us to provide more convenient areas for patient care.

Lun Ge: Once a student asked me why there was nothing new in designing hospitals. It is always in the shape of a Chinese calligraphic symbol "王".

Guoliang Chen: That's because the layout of a building represents its functional characteristics and the behavior patterns of its users. Looking at the symbol of "王", the part in the middle represents a concise and efficient connection between the medical department and the technical department, while the outpatient department in the front and the inpatient department in the back.

Naturally, we could develop many other layouts based on this principle. Only when fundamental changes happen to hospital operation and medical disciplines, will the architectural building plans alter drastically. Applicability is always the most important criterion for evaluating the quality of hospital building design.

I often give examples of a classroom's layout design. If we don't alter the pattern of "teachers giving lectures on a platform while students listening as the audience", the layout of a classroom won't be changed. Only when teachers step off the platform and sit together with their students in class, is there a possibility of changing classroom layouts.

Lun Ge: Which hospital in China do you think is relatively well designed, including the projects completed by your institute?

Guoliang Chen: Among the hospitals in a super-large scale of more than 200,000 m^2 that I've seen, Shenzhen Binhai Hospital is the one with well-established architectural frameworks and logical relations, including public spaces and transportation systems. However, I barely know its performance in practical operation, so I dare not to make any comments.

Lun Ge: It was inevitable that a plenty of contradictions occurred between the design orientation set up in early design and the mode of hospital operation in practice, since the University of Hong Kong took over this hospital. Regarding the rooms without a window, should a limit be set up on the amount, particularly in regard to the major rooms in a hospital?

Guoliang Chen: Firstly, I think it is necessary to classify hospitals. The scale of a hospital is of the greatest concern. Efficiency becomes more important than natural ventilation, when a hospital reaches to a certain large scale such as over 100,000 or 200,000 m^2. In order to promote natural ventilation, the distance between buildings would be increased and thus the distance for service flows to travel would be longer. Would it be worth to do so? For hospitals in a relatively small size (ranging from 80,000 to 90,000 m^2), it's possible to apply natural lighting and ventilation in most rooms as well as to ensure the efficiency of hospital operation.

Secondly, associated with hospital positioning, service intensification may be of top importance, especially in a hospital focusing on intensive care, emergency care, or high-end services with high technology. Take the brain surgery wards of Huashan Hospital in Shanghai as an example, it really doesn't matter if these wards are south or north facing since the most important thing is whether or not the design can help the nurses to get there as fast as possible. However, hospitals focusing on primary medical services are likely to put more emphasis on inpatient wards' natural lighting than nursing efficiency.

Moreover, when a hospital reaches to a certain large size, a dispersive layout with sub-centers can facilitate natural lighting and ventilation and thus reduce the impression of an oversized hospital. Therefore, the focuses of design may depend on the size and positioning of a hospital. So do the focuses of patients.

Lun Ge: Currently, joint design institutes including both domestic designers and those from foreign countries are popular. In most cases, oversea designers make a "shell" of the building or design the outlooking whereas domestic designers are responsible for building layout design. Hospitals and investors are frequently fooled by the outlooking and thus such kind of design always win design bids with ease.

Guoliang Chen: *In my view, it would be quite problematic if hospital design and construction were not carried out either from the inside to the outside or by integrating the interior and exterior design.*

In the case of a joint team composed of foreign and domestic designers, if the former built a "shell" and the latter took charge of the layout design, there would be a disaster. These groups couldn't run design simultaneously. Many years ago, a very famous hospital in China invited some oversea architects to design the appearance of their outpatient building. After they made decisions on the appearance of the building, domestic architects worked on the building's internal layout design. Although having rich experience designing hospitals, the domestic design institute still found it quite difficult to "fill in" certain functions in the "shell" without changing the building's appearance. After this building was built and in use, the design led to a lot of inconvenience for internal traffic flows.

So, I have the following thoughts: First of all, through these years of study and practice, our domestic architects are fully capable of doing a good job in modern hospital design. Secondly, in the case of inviting architects from foreign countries, I suggest that they should at least complete the design stage of design development. In other words, they are supposed to participate in the primary and secondary levels of design. In the end, what's necessary for us to bring in is their experience designing the entire hospital instead of only the appearance of buildings.

In the case of designing a general hospital providing primary medical services, you'd better not engaging architects from foreign countries. If your hospital has some special characteristics or plans to provide specialized medical services, you may engage some oversea architects with rich experience in hospital design.

To architects, the key to be successful in hospital design is closely related to how well we know about the positioning and operation mode of the hospital. A few years ago, we cooperated with an American design company who was good at hospital planning and design. Together, we completed a hospital design project funded by an American investment company. We were required to design in accordance with American modes of medical care. At the construction stage, most participants were directors and doctors hired from China and they put forward lots of suggestions modifying the design documents. The main point was to adjust the building layout according to the current medical operation modes in China. The entire design process was very disappointing and resulted in a huge waste of investment and time.

Fig. 4.4 Guoliang Chen and Lun Ge

We are now cooperating with several American designers in a project funded by a oversea investor, which aims at providing specialized medical services. A famous American medical consultant is responsible for the entire planning of medical technology. All future work is going to be based on and guided by this planning, including the design, construction, physician engagement, and management and operation modes. If we can insist on this method, I believe we can build a modern Chinese hospital that is fully in line with international standards.

As a result, I think the hospital party should be clear what kind of valuable ideas is expected to acquire from the oversea architects before hiring them. Do their ideas fit in with the status quo in China, in terms of operation and management?

Lun Ge: I had a very pleasant and enlightening chat with you today. I am happy that we have the same ideas which have inspired me a lot. Meanwhile, I benefited a lot from your penetrating opinions and insights as well. Thanks very much! (Fig. 4.4).

4.2 Obligations of Hospital Designers—An Exclusive Interview with Jian Gu

Time: December 2014—reiterated in May 2023
Location: The office of China IPPR International Engineering Co., Ltd., Beijing

Fig. 4.5 Jian Gu

Jian Gu: Chief architect of China IPPR International Engineering Co., Ltd., Deputy director of the institute for medical architectural design, first class registered architect, senior engineer.

Representative work: Foshan First People's Hospital, Guangdong Province; Foshan Women's and Children's Hospital, Guangdong Province (Fig. 4.5).

Key sentences:

1. *Therefore, if we need to sort out the technical processes for hospital design, we should figure out the flow relationships in accordance with the logic of medical services.*
2. *The brief of a project is critical as it is the basis guiding the design.*
3. *Interior design is not simply about decoration, but involves genuine 'functional designs' in the interior. It should be an extensive part of architectural design. Moreover, interior design influences service modes and involves certain concerns on psychology.*

Lun Ge: You have years of experience in research and design of medical buildings. How do you understand the complexity of hospital design?

Jian Gu: Hospital design is to a special and complicated type of architecture. It requires considerations on building design. Regarding the design of building properties, the most important one is to optimize user flows and workflows. It is kind of like a transportation building. There are three major "flows" in a hospital: persons, logistics, and information. Regarding the persons, it can be refined into patients, physicians and the support staff, and visitors. In the design of patient flows, cross-infection should be prevented, especially for pregnant women and sub-healthy people since they aren't patients. Meanwhile, we need to protect patient privacy and facilitate medical services and work efficiency via design. Relevant design items include 'green' paths for emergency services, surgeries, ICU care, etc. Focusing on the transportation and circulation of clean or contaminated items, logistics flows involve various routes and equipment. The information flows consist of the transmission and retrieval of various information and images. Moreover, various traffics should be separated, such as motor vehicles, freight vehicles, and pedestrians. *Therefore, if we need to sort out the technical processes for hospital design, we should figure out the flow relationships in accordance with the logic of medical services.*

Again, modern hospitals have certain similarities with commercial complexes. A hospital is to a great extent like a complex, which accommodates various types of activities, such as commercial activities, recreations, medical services including diagnosis, examination, treatment, rehabilitation, and health education, etc. All of these activities should be well organized.

Lun Ge: You led and participated in many significant hospital projects. What are the things that make an impression on you?

Jian Gu: Given the complexity of hospital architecture, our communication with the proprietors clients is certainly critical. *The most important thing is that mutual trust should be achieved between the two parties, since the preliminary stage of project.*

Hospital managers are familiar with medical services. During the communication with architects, they incline to use medical terminology. There are "gaps" between medical and architectural terminologies. It is necessary for architects to gain more knowledge on medical services in order to better explain their design ideas. Only by listening carefully to these managers and analyzing their needs, could architects better state their design ideas and reduce the "gaps". Communication is based on a common goal and mutual trust.

Communication needs compromise and concessions. The discussion on design is a process of weighing the pros and cons and making tradeoffs. Due to different expertise, perspectives, and circumstances, the two parties hold different views on design and attach different degrees of importance to certain things. It is necessary to put ourselves in their shoes. It is sensible for architects to silently sort thoughts in advance and know for sure what they shall adhere to as well as what they could compromise. With such sequence in mind, we could locate the key points, bottom lines, and master communication skills.

I always believe that the influence of clients on project success exceeds 50%. Previous projects proved this. It is undoubted that communication between the two parties would be more effective and successful if the clients had an open mind, tolerant attitude, and accurate judgement.

Taking myself as an example, an architect's career is actually very short. It has been more than 20 years since I started my first hospital project 'Foshan Hospital' with Dr. Huang in 1994 (Fig. 4.6). Since then, I have worked on six or seven projects in ten years. I think I can work on three or four more projects in another ten years, which is about ten hospitals in total. The chances of encountering an enlightened and like-minded client aren't high. When given such opportunities, we always want to break through ourselves and try to leave as less regret as possible. This is the driving force of my insistence on certain design ideas.

Lun Ge: I think that an architect's sense of accomplishment does not lie in the quantity of projects but the quality of design and good reputation. You adhere to

Fig. 4.6 Foshan First People's Hospital, Guangdong Province

your design principles. How do you handle the situation in which your clients know little about architecture?

Jian Gu: Sometimes architects need to insist on their opinions. Of course, you need to clarify the reasons why you want to "insist". "Insist" actually expresses a responsible attitude, and "insist" is also a kind of control. With specific perspectives and considerations, both sides need to collide for a consensus, so that both the spatial functions and architectural aesthetics are satisfying. Another thing is to double-check if there are better solutions for quality control.

For example, once we discussed with a client about how to deal with hospital traffics but ended up with no good solutions. Although the client accepted one of the suggested solutions, but I still felt that was not an ideal one. Finally, when I came up with a better idea, the joy in the client's eyes was a good return for my effort.

Lun Ge: It is very valuable for an architect to insist on his or her design ideas, so I really admire you for that. My feeling is that if you insist on your design ideas, the owner will respect you more. I often hear this from clients: the designers just draw whatever we tell them to draw, but with no ideas of their own. I think persistence comes from responsibility, self-confidence, and honesty. When the client finds that your insistence is right, he or she will accept your design. Of course, the ways to communicate is very important, as you just mentioned.

Jian Gu: When I was young, sometimes I was very anxious if encountering obstacles in communication. Now I won't.

Lun Ge: There is another problem when it comes to communicating with clients. For example, it is easy to communicate with the clients whose management are at a higher level. They can understand more about what you want to explain. How could you communicate with the clients who don't know much about design and construction?

Jian Gu: Perhaps, the third language exists. We live in the same society. So, there surely will be something both the clients and architects are familiar with. For instance, I had a chat with a friend with no architectural backgrounds. I took the process of using WeChat as an example to explain the size of a designer team, and my friend got it readily. Sometimes, using no terminologies could develop better communication.

Lun Ge: Then, could you please talk about the current problems of hospital design and construction in China?

Jian Gu: The first step of a project design and construction is planning, and after that comes the proposal. A study of project feasibility should be taken as the precondition of project planning and require careful analysis of the project advantages and disadvantages. Without an in-depth planning, the project would be wrong from the beginning. How could we achieve a convincing outcome? What's more, the team of planning shall recruit hospital administrators, medical experts, policy advisors, architects, planners, and financial consultants.

We should start a hospital project by analyzing its medical strategies and resources, instead of merely presenting a design drawing. A project proposal, a report on project feasibility, and a project brief should be formulated in accordance with the analysis. Today's procedures of hospital design and construction in China lack a step of rational planning. I read an American report on project feasibility. Sixty percent of the report was focused on research charts, including near-term and long-term goals, traffic systems of motor vehicles and pedestrians, flexible building functions, available municipal resources, and green hospital strategies. Moreover, detailed traffic organization and temporary buildings for construction were introduced.

Another key point of planning is prospective thinking. It doesn't mean a prediction on patient increases, but refers to the consideration of future policies and advanced methods of medical care. Hospitals in China are greatly affected by policies. Medical reforms and the new orientation of hospital development, including the grading of hospitals, the methods of insurance reimbursement, diagnostic and treatment methods, influence the design and construction of hospitals. The government of Beijing introduced a new plan of medical reform, which updated the orientation and tasks of public hospitals via a "two-way referral" system. In the new system, key hospitals are no longer allowed to construct new outpatient buildings. Fundamental medical care would be decentralized to primary hospitals. Furthermore, an integrated system of medical treatment and rehabilitation would be developed to increase the utilization of primary hospitals and improve the quality of care. Moreover, the number of outpatients keep increasing in large general hospitals. All these variables will affect hospital layout designs and the area proportion of functional departments. We won't know when the new policies would be launched. However, the trend is inevitable and we need to be prepared.

The second problem is irrationality. In recent hospital projects, a popular slogan "keeping the current for 50 years" have been interpreted. It's often the case that a hospital lobby is in a size of 1000 m^2. The lobby, however, is empty with no functional design at all. I visited a Japanese hospital. Its lobby was just about 400–500 m^2 and had multiple functional areas such as a reception, a registration, self-help medical service areas, catering service areas, and waiting areas. What's more, traffic flows in the lobby were clear. Although the size of a hospital in Japan is relatively small, the rational way of design is worthy of learning. Furthermore, regarding the hospitals in China, it seems that a large entry square with a width of more than 70 m has been a required design factor. Typically, these squares are far away from public transportations. A few research has been done regarding the vehicles and pedestrians in the squares in terms of safety, accessibility, and traffic efficiency.

Moreover, it looks like 'main streets' is necessary for both large and small hospitals. However, the streets are generally addressed as wide corridors for transportation. As far as I'm concerned, in a large hospital, the main street should be an integrated space containing transportation, commercial and recreational functions. It should be a place without medical atmosphere where people would like to stay for a while, rather than a simple linear corridor to pass through. Still, we have to be wary of the over-commercial tendency of hospital design. In some hospitals, public spaces, building appearances, and even building structures have been over-designed and exaggerated.

It is true that we should learn from other types of public buildings in terms of design. However, buildings have varied characteristics and natures. We shall advocate plain and serene hospital designs and focuses more on the functions.

The third problem is that Chinese designers lack self-confidence. Normally if someone were unconfident, they typically need additional things to decorate and enhance their so-called value, just like many go crazy to grab LV bags, pile titles on their business cards and so on. The design competitions initiated by some hospitals in China are very 'oversea worship': they invite many famous oversea design companies and even refuse local design institutes to participate. These hospitals spent a lot of money on design and received a bunch of shallow design concepts. Oversea design companies also know that and thus make 'fierce' or exaggerated designs in order to be competitive in these competitions.

The value of oversea designers doesn't lie in their design renderings. It is in the design process ideas and the ways to refine designs. If possible, an oversea design company could be entrusted throughout the entire project from the first stage to the last. So that their advanced international design ideas could be put into practice in China. President Xi also emphasizes on the importance of cultural confidence and self-esteem. This is especially the case of hospital design.

The fourth problem is that we have done little research on design and paid no attention to design details. Few designs in China have been done under the guidance of theories or based on empirical research data. Evidence-based design in America has been conducted for several decades and becomes the foundation of hospital design. In most cases, our designers learn from experiences and design customs. Experiences and customs have their inherent inevitability, but not all of them are reasonable. Hospitals are developed rapidly. Hospital buildings need updates and our designers need the supports from research. Moreover, we should pay more attention to design details. How to promote patient care via detailed design? How to reduce the workloads of medical staff? The distance that a nurse in an inpatient department travels daily may up to 7 km. I made a rough calculation. If the grid of columns in an inpatient building was adjusted from 7.2 m to 8.1, the nurse would need to walk one more kilometer per day. Since then, I couldn't help wondering: How to consider these factors in design? What is an appropriate way to handle these concerns? Hence, we deeply hope that you would do some in-depth research on these topics.

Lun Ge: Speaking of research, I attended international conferences every year. What impressed me the most was that both research institutions and design companies conducted research. None of the design companies merely focused on design in their presentations. Generally, their design works were based on research. The research conducted by our team focuses on preliminary design, methods, and contents, such as the formulation of a project brief. What is your advice in this regard?

Jian Gu: *The brief of a project is critical as it is the basis guiding the design.* However, project briefs that we received were often of no value of guidance for design. Sometimes, a brief has only two pages of empty conventional talks or slogans. It contains little professional information or specific needs of the project. Indeed, a project brief is more than a formality. Each hospital has its own peculiarities. A brief should represent the characteristics. Only via a project brief, could designers know the special requirements of the hospital and make customized design accordingly. Once we participated in a project bidding abroad. The project brief had 300 pages, including numerous bubble charts of functional analysis and room lists. These were quite illustrative and truly helpful to design. The notes in the brief were clear and specific on technical issues, instead of being an empty talk, like "conforming to the policies, keeping current for at least 50 years, putting patients first", and so on.

Lun Ge: Do you think technical design should be viewed as part of a hospital design? Should it be reviewed? If so, at which design stage? Before or after the reviews on the design development or construction documents?

Jian Gu: *Technical design is very important. The tables and bubble diagrams included in a project brief are developed via technical consultations.* Once a conceptual design based on the project brief were created, the mode of hospital management would be determined too. I think technical design should be reviewed before the review of design-development documents, because it is still part of the scheme design.

Lun Ge: Also, we had discussions in regard to the contents of design and the time conflicts of interior design with architectural design. What's your opinion on the conflicts?

Jian Gu: Many clients think that interior design equals to 'decoration', which is a misunderstanding. However, that's why many hospitals separate interior design projects from architectural design. *Interior design is not simply about decoration, but involves genuine 'functional designs' in the interior. It should be an extensive part of architectural design.*

A hospital provides services to patients and treats them as 'clients.' Interior design emphasizes on providing not only good user experience but also healing environments to facilitate rather than harm patient recovery. Regarding the environments for rehabilitation, environmental functions play a prominent role. In this regard, interior design is an extension of architectural design. For instance, waiting areas for registration are different from those for medical treatment in terms of the degree of urgency and the duration of expected waiting time. Thus, the demands for interpersonal communications in these areas are different. So do the requirements for furniture, spatial layouts, and environmental atmosphere. More attention should be paid to patients' experience during the entire process of a hospital visit. Many hospitals address interior design as an independent project. However, systematic thinking and the ideology of integrated functional designs are needed. An important goal of spatial design is to facilitate environmental functions. Design can only be made after meeting the functional needs such as the design of supportive items encouraging people to stay in the places where they need to stay. What kind of spatial forms and atmosphere should be developed? How to display the furniture? How to plan the areas for recreation, waiting, catering services, or other public activities? *Moreover, interior design influences service modes and involves certain concerns on psychology.* The requirements for creating appropriate environmental atmosphere and interior colors are certainly different between children's hospitals and general hospitals. A mode of VIP services also results in specific spatial design requirements (Fig. 4.7).

We need to consider how to maintain users' sense of social participation via interior design. Hospital buildings are public buildings that provide medical services. Modern hospitals are small societies in which social activities keep increasing, such as catering, shopping, reading, health education, recreation, and religious activities. People should not be deprived of social connections in hospitals. They concern whether their medical needs are met and whether their dignity and privacy are respected. Hospital users should be encouraged to join the design of healing environments for their hospitals. It has been leading the development of modern hospitals and part of humanity. For instance, in some hospitals aboard, displaying patients' photography works and paintings on hospital walls are aimed at maintaining the sense of social participation.

Lun Ge: Could you share your thoughts on the post occupancy evaluation (POE) of hospital buildings?

Fig. 4.7 Foshan Women's and Children's Hospital, Guangdong Province

Jian Gu: I think that POE should be the last step of project planning and design. The aims are to investigate to what extent that a built project follows its targets, preliminary planning, and final design. Based on the investigation some universal laws of design can be gained. Then we know what needs to be improved and how to pave the way for subsequent projects.

Lun Ge: I'm glad that you've extended the contents and meanings of POE that we've been working on.

Jian Gu: Of course, POE should first be based on a rational plan. Without a question for evaluation, the answers to POE will naturally be various. What would do you take as a benchmark for POE?

Lun Ge: Through on-site surveys, we found that many hospital departments lacked a lot of functional rooms. This phenomenon is associated with the changing requirements and personnels during the long period of hospital construction. I would like to hear your comments on it.

Jian Gu: We should consider it within a timeline. It's true that the duration of a hospital construction project is typically very long. From a project approval to the completion of buildings, drastic changes could happen. For example, in one of my projects, the number of outpatients per day was expected to be 3000 at the time receiving the project approval but actually reached up to 12,000 in two years since the hospital was in operation. Similarly, in another project, the designed consumption of steam was far beyond the actual usage. The reason was that the proposed laundry services was moved out of the hospital campus. All these factors need to be taken into consideration during POE.

Lun Ge: Here is the last question: What's your advice on managing the quality of hospital design?

Jian Gu: As I mentioned before, figuring out the procedures is the first step. We need to handle macro issues first, then medium ones, and micro issues the last. The ways of thinking regarding each procedure are the same. This helps us to avoid mistakes.

Secondly, we shall plan and design hospitals rationally and logically, like solving mathematic problems. If so, the results of design would be more reasonable.

The third one is about functions. We should be aware that functions are ubiquitous and more attention should be paid to the specific functions of a hospital. Hospital buildings are not for display but practice.

Fourthly, each hospital should be designed in accordance of its own positioning and characteristics. The portfolio of a design institute should in no way contain design replicas or things like the samples issued by a rendering company.

The last point is to let the professionals do their jobs. Both designers and hospital clients have their own perspectives and responsibilities. It's necessary for them to cooperate and coordinate with each other, with their professional knowledge and advantages. We should pull together so that one plus one will make more than two.

Lun Ge: Thanks for your wonderful sharing (Fig. 4.8).

Fig. 4.8 Jian Gu and Lun Ge

4.3 Medical Architects and Challenges in the Era of Transition—An Exclusive Interview with Dr. Xiqiu Huang

Time: December 2012—reiterated in July 2023
Location: The office of China IPPR International Engineering Co., Ltd., Beijing

Xiqiu Huang: Famous medical architect in China, Winner of the 6th "Liang Sicheng Architecture Award".

Representative work: Foshan First People's Hospital, Guangdong province; Beijing Xiaotangshan Hospital, Beijing (Fig. 4.9).

Key sentences:

1. *Modern hospital design and construction in China has been developing more and more rapidly after the Reform and Open policy issued in China. However, our architects are not well prepared in terms of understanding the policies, design concepts and methods of designing hospitals.*
2. *Regarding the procedures, what we are used to making decisions on the scale and budget before conducting project feasibility study. This sequence causes many problems.*
3. *Due to insufficient development of community-level medical services, it is hard to create a completed hierarchical healthcare system and thus leads to the phenomenon of overcrowded at large hospitals in big cities.*
4. *At the turning point of national development, there are insufficient connections among disciplines, insufficient preliminary planning, and ambiguous positioning of a hospital project. Rational design and development of hospitals is necessary.*

Lun Ge: Dr. Huang, you are a recognized leader in the field of medical architecture in China. You have witnessed the development of Chinese hospitals since the

Fig. 4.9 Xiqiu Huang

1980s. Would you share your thoughts and experiences regarding the development of hospitals in China?

Xiqiu Huang: Hospital construction in China has been developing rapidly since the national reform and opening up. *However, as a medical architect, I think that we aren't prepared for the fast development.* My colleagues and I completed many projects, but none of them was likely to fully meet the social demands in its true essence, especially the actual needs of our populations. During the current period of national transition and upgrading, it's inevitable that more rural populations flood into cities, leading to new-types of urbanization. Under such a circumstance, there would be shortages of educational and medical resources. Moreover, our existing systems of healthcare as well as the associated facilities and planning couldn't meet the current demands.

Despite of the rapid development, we lack both pre-design research and post occupancy evaluation (POE) of our projects. A lot of work should be done to keep pace with the rapid development. Many decisions on a project were based on thoughts regarding its local areas or even subjective ideas or emotional judgment, including those on project location, number of beds, functional requirement, target populations, and references on expected changes of disease spectrums. Particularly, population movement patterns[1] keep changing and are no longer the same as those in the initial stages of development. Big cities are under stricter management and control whereas some small cities develop faster. This raises the question: How would health care systems do to support the changes and needs? In my view, it's necessary to invite credited experts in public health to support this trend and medical planning and share their services in both urban and rural areas. However, the interdisciplinary coordination between architects and other professionals in these fields still has a long way to go.

In recent years, general hospitals have been developed fast. However, specialty healthcare facilities such as children's hospitals, cancer hospitals, mental hospitals, and geriatric hospitals are left behind. China is still short in nursing services for pregnant women, cancer patients, wounded or handicapped persons, and seniors. These require medical architects to engage in meticulous studies on their needs, in order to improve the quality of design for various types of hospitals.

Today, we have more chances to collaborations with oversea designers and could learn from their experiences both in developing design concepts and methods. Meanwhile, we shall carefully apply advanced oversea experiences within the context of China. For instance, the strategy of designing all patient rooms as single private rooms has been advocated in America because of high levels of usability, patient satisfaction, and reduced infection rates. I agree on the suggested advantages of single-patient rooms but there are practical difficulties of implementing this strategy in today's China. It's not practical. China has a large population and still short of hospitals. The current healthcare systems aren't strong enough to meet all the demands. Therefore single-patient rooms are only available for limited groups of people.

Healthcare systems in urban areas may be generally better developed than those in rural areas where the single-patient room strategy may not be applicable. From

Fig. 4.10 Foshan First People's Hospital, Guangdong Province

this point of view, China is far behind in comparison of the developed countries. We benefit from learning oversea experiences but shall be aware of cross-national differences. Certainly, there is nothing wrong paying attention to advanced strategies and we may use them when our hospitals have been improved to certain levels in future (Fig. 4.10).

Architects, especially medical architects, should consider how to meet the needs of our society. Many hospitals over value the importance of their buildings' appearance. They also make the public areas inappropriately big in size. To my knowledge, even developed countries don't make their hospital areas in such a big size. Someone even suggest that corridors in an inpatient department should be 4-m wide or wider. It is too much.

Lun Ge: You shared thoughts regarding the healthcare systems in China at a macro level. What's your opinions on the procedures of hospital design and construction?

Xiqiu Huang: The Ministry of Construction (MOC) has published detailed regulations regarding project construction procedures in the past few years. However, in some projects, the reports on project feasibility were conducted after the decisions on project sizes and budget were made. If the sizes and budget proposed in the reports are different from the decisions, they will not be accepted and thus the investor would not proceed the design and construction. For instance, in one hospital project that design was done by my team, the local government decided the building size and budget and then we were asked to build a hospital in that size and within that budget. We were supposed to develop a report on project feasibility, but the size and budget couldn't meet. I informed them directly that our estimates on the budget was higher, which was carefully calculated based on the expected size, construction standards, and local quotas. However, the authorities insisted that the size couldn't be reduced and the budget cannot be exceeded. Later, we had to reduce the figure by moving a well-planned budget for the inner decoration of operating rooms into another budget on specialties.

In recent years, hospitals in China get bigger and bigger but still cannot meet the needs of the society.

National policies intend to develop hierarchical healthcare systems and distribute healthcare services at multiple levels. Community-level hospitals are expected to take care of common diseases and thus many patients don't need to visit large hospitals. However, we frequently see children and older adults travelling to see doctors in large hospitals since community-level hospitals aren't able to provide high quality services.

For instance, 50% or 60% of the patients visiting the hospitals in Beijing aren't Beijing residents. In no ways could we keep them away since they have no access to high-quality healthcare services back in their hometowns. If healthcare resources were distributed more equally and high-quality hospitals were developed in every province, hospitals in the capital may not be overcrowded. People living in big cities also tend to visit large hospitals to see famous doctors. High-level strategic healthcare planning is needed in order to better distribute patients and develop referral and reimbursement systems.

Lun Ge: What are your thoughts regarding the communication between architects and hospital investors? You mentioned that architects weren't fully prepared. Is this a problem?

Xiqiu Huang: Medical architects should make themselves prepared with professional knowledge and skills and be open to suggestions from their hospital clients. However, I noticed that architects may have attitude problems: Firstly, some take a passive attitude in learning medical knowledge since it goes beyond the field of architectural design and are stubborn on their design ideas without sufficient consideration regarding the needs of patients and hospital staff; this won't lead to a high quality design since there is too much focus on the architecture itself; Secondly, some have no idea on how to design but blindly follow the directions suggested by their hospital clients and accept whatever the clients say; the clients won't be satisfactory either. A medical architect should provide professional design services to his or her clients. How come the design is totally based on clients' opinions? I would like to suggest medical architects to improve their knowledge on medical care and communication skills in order to be well prepared for hospital design.

Lun Ge: With years of hospital design practice, do you still have problems in the language translation between architecture and medical care?

Xiqiu Huang: Of course, I do. Because of different professional backgrounds such as medicine or architecture, architects should carefully find design solutions to meet hospital needs and translate the solutions into architectural drawings. The hospital designs and layouts should be able to express their design responses in a professional way. It requires a lot of patience. Architects should communicate well with their hospital clients.

Lun Ge: At the early stage of design, practical architects may not have direct communication with their hospital clients. A third-party consulting company, similar to those in foreign countries, has been suggested to connect architects and hospitals. Do you think it is necessary?

Xiqiu Huang: It depends on how professional the architects are. If they have never designed hospital buildings, it's suggested to cooperate with a consulting company. If they are prepared with professional skills in both consultancy and hospital design, then it is not necessary to work with a consulting company. It depends on individual cases.

Lun Ge: You mentioned that interdisciplinary connections should be encouraged in China, to promote pre-design research, project definition and positioning, and the growing sizes of hospitals. Any other comments?

Xiqiu Huang: It seems that we haven't done very well in the rational development of hospitals.

Lun Ge: This situation has started changing. In recent two years, it has been advocated to build a hospital as an aircraft carrier.

Xiqiu Huang: It has emphasized on constructing taller and bigger hospital buildings in order to be a 'world-class' hospital. Every hospital intended to do that regardless of its location and economic conditions. It's no longer advocated. As far as I know, many advanced hospital buildings are built in the Middle East and other rich countries. In recent years, we have seen the resurgence of 'big' hospitals in China. However, I saw several new hospitals built in U.S. and other countries were quite properly and rationally constructed.

Lun Ge: The United States has a quality control mechanism in place but China doesn't.

Xiqiu Huang: Hospital construction is relatively rational. The quality control mechanism does exist in China too. If the scale of a project went beyond the investment, the authorities wouldn't approve it. Project efficiency will be valued when it put into operation.

Lun Ge: Our pre-design research involves hospital project briefs and the processes of medical services. Any thoughts on it?

Xiqiu Huang: The brief of a hospital project in China isn't in detail. We need to clarify the functions of a hospital and then conduct a programming of the functions. However, it's relatively difficult for architects to propose a hospital with themselves since there are many public health factors involving in hospital design. Hence, interdisciplinary collaborations and the involvement with high levels of hospital management are required for longitudinal studies leading to more applicable results (Fig. 4.11).

Lun Ge: Customized design is critical to hospital buildings. Your company has done excellent jobs in this regard.

Xiqiu Huang: Compared with others companies, our company conducted more hospital designs and thus got more opportunities to learn. Medical architects are

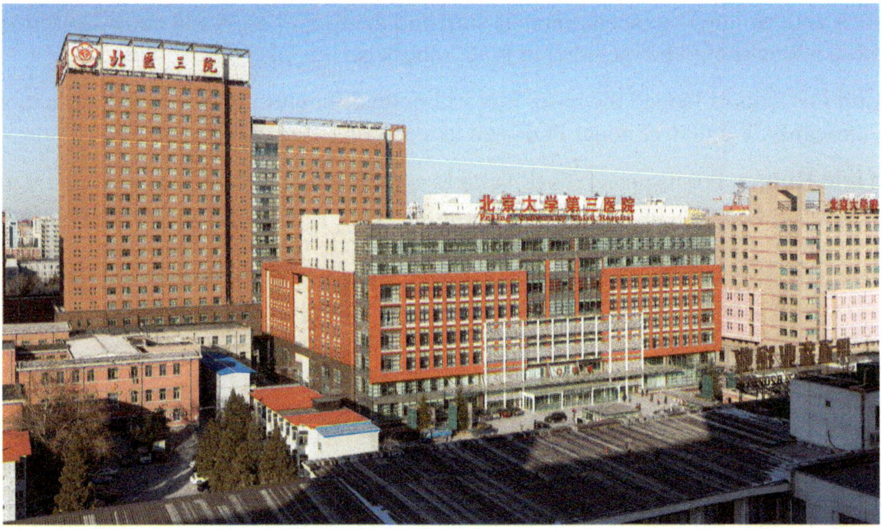

Fig. 4.11 Beijing University Third Hospital, Beijing

required to continuous learn and keep updating their knowledge on new equipment, the processes of medical services, and treatment methods. That's why I suggest architects to establish usable channels for communication with doctors and nurses in order to learn from them. Knowledge changes by day.

Lun Ge: We conducted research on the POE of hospital projects and developed the evaluation tool SHAPE. It intends to evaluate a hospital building from five perspectives: safety, humanity, functions and spaces, physical environment, and efficiency. We would like to hear your thoughts on hospital POE in China. Is it difficult to conduct? Should it be included in the procedures of a hospital design project? Given the missing of POE in hospital development, is it necessary to include it in the required project procedures and carry in out?

Xiqiu Huang: Regarding POE, we should first have an in-depth understanding of it. The workloads of POE are heavy. Each hospital has some distinctive situations. Take the factor of surrounding traffics as an example, the situations of Beijing Union Medical College Hospital and the 3rd Hospital of Beijing University are far from ideal and these are associated with their historical backgrounds. To solve the overcrowding problems of urban hospitals, relocation them to suburbs has been suggested but it would cause new problems. In our discussion with the official staff from the Ministry of Health, we suggested to move a large hospital to a suburban location. The question is if we moved it, the hospital staff and their families would still live in the prior place and thus the new hospital would be over twenty miles away from their home. Many shuttle buses should be provided for their daily commute. Another case is if we built a center for disease control in the suburbs, we should build associated amenity facilities such as dormitories to accommodate staff and their families. A full consideration of these issues should be included in POE, otherwise it is likely to be biased.

Lun Ge: This is indeed the case. It is still quite difficult to carry out the evaluation thoroughly. We may conduct POE within a certain range.

Xiqiu Huang: For instance, it's difficult to appropriately evaluate the size of a room. Spaces are typically bigger in hospitals with generous investments. If staff members were asked about the size of a locker room, they definitely want it to be spacious and they have various reasons. It is not easy to evaluate these factors. A head nurse may have more experience and thorough considerations regarding hospital design. If we ask nurses regarding their needs in hospital, they are more likely to focus on their workplaces. Comparison and analysis are needed since the demands and opinions differ from one another.

There are many factors influencing POE. It's necessary to collect samples for research. Evidence-based design requires data analysis and comparison for rational design solutions. We should never jump to a conclusion and then make inferences based on that.

Lun Ge: We met problems during POE. Like what you said, how to conduct POE is critical. Whether or not a hospital cooperates with POE and whether or not the government values its importance are important. We hope the government can make POE a mandatory requirement for hospital development.

Xiqiu Huang: Strictly saying, it should be in that way. Similar to the trend of green hospitals, the application of POE needs to be promoted by both the Ministry of Construction and the Ministry of Health. There are barriers to POE surveys. For instance, hospitals may not allow us to visit some of their departments due to the large number of visitors.

Lun Ge: It's true. We encountered various difficulties in survey, such as the sample sizes were small. We hope to identify the commonality in hospital design by conducting case studies. Should the government conduct hospital POE and make POE a regular procedure required for hospital projects?

Xiqiu Huang: It is hard to tell at this time. We need more experience and nationwide attention on POE in order to promote it. To my knowledge, an oversea hospital may contact a consulting company to schedule a POE when it's necessary. It's voluntary rather than mandatory.

Lun Ge: Building quality has received a lot of attention. Do you think there should be a criterion for evaluating the quality of hospital buildings, similar to that for rating the star level of a hotel? In this way, quality can be linked to investment. Do you think it's necessary for hospitals to establish such a quality evaluation system?

Xiqiu Huang: Hotel evaluation systems are designed for hotels, including both the software and hardware evaluations. However, regarding hospitals, we still work on the basic classification of hospital ratings[2] and POE could be the next topic. How could we evaluate one single factor (the buildings) of a hospital without its rating on

Fig. 4.12 Xiqiu Huang and Lun Ge

medical services? Evaluating a hospital should majorly focus on the quality and quantity of medical doctors in the hospital. The quality of health services should be the most important factor of hospital evaluation and the factor of building quality can go after it.

Given the existing standards for hospital rating, we may propose supplement standards for hospital building rating. It would be meaningless if making the independent standards for hospital buildings and difficult to carry it out as well.

Lun Ge: Thanks very much for your sharing (Fig. 4.12).

4.4 Comprehensive Hospital Design for Humanistic Care—An Exclusive Interview of Jianmin Meng

Date: January 2014—reiterated in May 2023
Location: MENG Design and Research Studio, Shenzhen

Jianmin Meng: Member of the Chinese Academy of Engineering; Chief Architect of Shenzhen General Institute of Architectural Design and Research Co., Ltd.; Distinguished Professor of Shenzhen University; Recipient of the 7th Liang Sicheng Architecture Award.

Fig. 4.13 Jianmin Meng

Representative work: Shenzhen Hospital affiliated with the University of Hong Kong, Guangdong Province; Zhangjiagang First People's Hospital, Jiangsu Province (Fig. 4.13).

Key sentences:

1. *Without a good planning, the design would be rigid or merely repeating the precedents without any improvement. One of our clients frankly criticized our design institute for repetitive design mistakes. He said that our designers kept carrying the same mistakes from one project to the other. This really upset me. Since then, I often refer it to warn our designers. Actually, this is a common problem in many design institutes.*
2. *First of all, the most fundamental requirement is to meet the needs for hospital functions, including medical technologies and service processes. Secondly, we have been advocating the concept of all-round care, which embodies the ideology of humanity. In other words, we care all the user groups in a hospital instead of merely the patients. The third point involves the idea of green and ecological development (ventilation, natural lighting, etc.). The fourth one is about technologies and costs. We suggest budgeted design and economic efficiency instead of blindly using expensive products or materials. Fifthly, we should incorporate the ideas of urban and architectural aesthetics in design.*

Lun Ge: Based on your years of practical experience in the field of hospital design, what do you think about the missing procedures in design or the existing problems? The conclusion and finding from our years of research suggest to promote the procedures of preliminary design and the post occupancy evaluation (POE).

Jianmin Meng: Your team work remedies two important design procedures currently missing in hospital design: preliminary planning and POE. Without a good planning, the design would be rigid or merely repeating the precedents without any improvement. One of our clients frankly criticized our design institute for repetitive design mistakes. He said that our designers kept carrying the same mistakes from one project to the other. This really upset me. Since then, I often refer it to warn our designers. Actually, this is a common problem in many design institutes.

Designers keep making the same mistakes and producing up-to-standard design works, instead of excellent ones. Although these designs may pass design reviews but the quality of design is far from excellent. Why? That's because many design institutes use the basic national construction codes and project briefs as the only basis and guidelines for design and focus too much on making profits. They would deem their duties completely performed once a contract got signed and the design documents passed review.

Lun Ge: What you said exactly coincided with the purpose of our POE research. POE should be a required procedure in design.

Jianmin Meng: Yes, it's necessary to conduct POE and summarize both the success and failure of each project, including the experiences and the lessons. This is a way of learning and improving. Nowadays, many design institutes only concern how to complete design tasks and get paid. Although some may pay attention to their built projects but there are no organized efforts to systematically summarize design practice and experience. This is a common problem in the design market of China. In fact, preliminary planning and POE are the best approaches to avoid repeating mistakes or producing low-quality design. So, what you are doing is very meaningful.

Lun Ge: Based on your hospital design experience collected in the past decade, could you give suggestions and comments on SHAPE, a POE system we recently developed for hospital buildings?

Jianmin Meng: It's pretty good. Actually, you set up an evaluation system.

Lun Ge: On the other hand, it can serve as the standards for quality control. So, this is one system with two aspects. It can be used to evaluate whether the quality of design or the hospital functions in the built project meet the project requirements.

Jianmin Meng: Have you reached any conclusions after using SHAPE to evaluate hospital projects?

Lun Ge: Yes. We do have conclusions. For instance, what is the appropriate size of a column grid for nursing unit design? If there were 45 patient beds, the floor area of 1500 m^2 would be insufficient and cause a short of auxiliary rooms. On earth, which size is appropriate for a nursing unit? We found the answers. Our findings are based on evidence, which helps to identify the key items of design for quality control.

Since we have completed many hospital POEs, we realize deeply that there are unreasonable big wastes in hospital projects. Many projects looked great from the outside and were built with expensive materials such as high-grade stones. However, things were very different once we probed into the process of medical services and had conversations with staff members. Deeply, we realize that many design mistakes, especially those associated with the fundamental items of design, such as spatial functions, could be avoided or reduced if the preliminary planning were performed properly.

Jianmin Meng: Is it possible to correct the mistakes once identified them in POE?

Lun Ge: Our POE focused on the hospitals that had been occupied for two years or more. Generally, after five years of use, building modifications are inevitable for a hospital. With a POE report, the hospital should be clear regarding existing problems and ready for modifications. POE is of guiding significance. POE also contributes to the establishment of a database that includes both project successes and lessons learned. It can be used as a reference for future hospital development so the same mistakes can be avoided.

Jianmin Meng: It's true. POE plays a fundamental role in guiding the improvement of design quality.

Lun Ge: Generally, design-development documents are sent to a design-review company for design review. Should the technical design of functional flows and equipment systems be reviewed before that? Also, should these specialty reviews be added to both design-development reviews and construction-documents reviews?

Jianmin Meng: The answer is yes. Currently, these reviews are carried out in the form of an expert workshop. However, sometimes the workshop may not pay enough attention to the technical process of medical services or functions.

Several years ago, a hospital in Shenzhen invited experts in the field of geotechnical and structural engineering to review hospital design. These experts knew little about hospital design and thus their comments didn't help. These experts were selected by a lucky drew from a group of over 100 experts. Their areas of expertise and specialty were individually different. So, the way of randomly drawing experts to review hospital design was quite inappropriate. In my view, we should set up professional review committees for hospital projects rather than randomly inviting experts to review.

For instance, the international design competitions for key projects usually have two steps of review. The first step is focused on technical review and then comes the experts review on design. For the reviews of hospital design, the first step is missing. No matter there were five or ten proposals, a technical review committee should identify possible mistakes in the designs of medical circulations or functional layouts, and develop evaluation reports, which may serve as references for experts' review. Only in this way, will the experts review be evidence-based and supported by health authorities.

Lun Ge: I hope the findings of our POE research can be verified and distributed via professional training.

Jianmin Meng: This is very important. Regular professional training should help to improve the overall competence of a designer and emphasize the importance of preliminary planning. Preliminary planning or preliminary design is the first phase of design preparation just like the prenatal care for a good birth.

Lun Ge: It's true. Every time when I work on preliminary planning for a hospital project, I emphasize that hospital consultants are like medical doctors. Firstly, we conduct exams and diagnoses to identify diseases. Then, we provide a therapeutic proposal with prescriptions for step-by-step treatments. Eventually, the project can be started with a healthy beginning.

Jianmin Meng: Preliminary planning at the early stage of design is like the prenatal care for a healthy baby; POE is like the prescriptions for disease treatments. Of hospital renovations, the treatment-like projects can be applied at three levels: cost-effective renovation with a limited budget, upgrading renovation or reconstruction, and intensive reconstruction. For hospitals with limited budget, it is very important to know that a slight modification would make a big difference. Some hospitals are financially sound and can afford to reconstruct their buildings in a thorough way. Variations and targeted strategies will lead the trends of hospital design market in the future.

Lun Ge: We updated the classification of hospital areas from seven departments into ten systems by medical function and spatial organization. What is the key difference between them? The seven departments are defined from the perspective of hospital management, whereas the ten systems are defined according to hospitals' medical functions and spatial organization.

Many experts think the 2008 version of *National Construction Standards for General Hospitals*, which introduces the concept of 7-department classification, has fallen behind. What's your opinion on this?

Jianmin Meng: The method of organizing and classifying functional medical areas by system is quite reasonable. It is comprehensive to include all the systems. What is the specialty of public spaces in these ten systems?

Lun Ge: It includes the spaces such as convenient stores, supermarkets, and catering spaces for food and drink, where humanistic services may be provided. If following the 7-department concept, we would have no ideas to classify these public spaces. It just doesn't feel right to assign them to any of the departments.

Jianmin Meng: These spaces are objective existences and the humanistic services should be reinforced in the future of hospital designs. They should be presented as an independent system. The new classification is scientific.

Lun Ge: I would like to ask, what are the necessary elements for an outstanding hospital design?

Jianmin Meng: First of all, the most fundamental requirement is to meet the needs for hospital functions, including medical technologies and service processes. Secondly, we have been advocating the concept of all-round care, which embodies the ideology of humanity. In other words, we care all the user groups in a hospital instead of merely the patients. The third point involves the idea of green and ecological development (ventilation, natural lighting, etc.). The fourth one is about technologies and costs. We suggest budgeted design and economic efficiency instead of blindly using expensive products or materials. Fifthly, we should incorporate the ideas of urban and architectural aesthetics in design. Our team designed many cultural buildings. *In our hospital projects, we not only pay special attentions to the medical particularities* *of a hospital, but also leverage on our advantages in creating architectural forms, environments, and urban spaces.*

Lun Ge: Do you think there are any good examples of hospital design in China?

Jianmin Meng: Of our design, the satisfied projects include the University of Hong Kong Shenzhen Hospital (Fig. 4.14), Zhangjiagang First People's Hospital (Fig. 4.15), the Second Affiliated Hospital of Anhui Medical University (with very limited budget), and the First People's Hospital affiliate with Jilin University.

Lun Ge: What do you think about the ways that oversea design teams collaborate with domestic design teams?

Jianmin Meng: A few years ago, in the bidding for a hospital project in Shenzhen, we once co-operated with an oversea design agency. In the beginning, one oversea architect was working on the concept design and our team was working on design development. However, when the schedule was half over, we realized that the differences between the oversea design scheme and the actual requests of the client was too far to win the bidding. We immediately decided to make another design proposal on our own, which finally won the bidding. What I want to indicate with this example is that when an oversea agency participates in domestic hospital design, the oversea company needs to meet two prerequisites: one is to be dedicated, and the other is to have a good understanding of the medical markets in China and the demands from

Fig. 4.14 The University of Hong Kong Shenzhen Hospital, Guangdong Province

Fig. 4.15 Zhangjiagang First People's Hospital, Jiangsu Province

the clients. That's the basis for collaboration and develops the possibility for creating a high-quality design.

Lun Ge: Thanks for your wonderful sharing of experience and insight (Fig. 4.16).

Fig. 4.16 Jianmin Meng and Lun Ge

4.5 As Bold as a "Dinosaur" in This Commercial Age—An Exclusive Interview with Xi Zhu

Time: January 2014—reiterated in May 2023
Location: A lovely and sunny restaurant, Shenzhen

Xi Zhu: Vice president in charge of the design of Bouygues Construction Group (France); BMP, Shenzhen Baipeng Architectural Design Co., Ltd.; Senior engineer who receives special allowance from the government; First-class registered architect.

Representative works: the First Hospital affiliated with Zhengzhou University, Henan Province; the First Hospital affiliated with Guangdong Medical University, National Center for Respiratory Medicine (Fig. 4.17).

Key sentences:

1. *In a hospital project, initial design review is a key step in design assessment, which includes comprehensive assessment of architecture design and technological design. Design institutes revise and improve their designs according to reviewers' comments.*
2. *Construction drawing design, including architecture design, interior design, and specialty projects, need to be done simultaneously since design factors may overlap and interrelated with each other. An executive architect is responsible to administrate and coordinate design as a whole to prevent design factors from separation.*
3. *The "7 Departments" principle in the 2008 Building Codes is vague, whereas projects vary dramatically and thus classifying is difficult. The idea of classifying by system, provides a detailed and suitable method. In a hospital, the proportion of an individual system depends on the hospital. We could gain experience from other fields. For example, we learn the planning of construction from real estate and project classification from hotels.*

Fig. 4.17 Xi Zhu

About tendering

Lun Ge: During the bidding for a hospital project, I often came across the situations that almost all the design companies put their efforts on eye-catching rendering instead of program organization. After the 'shell' of a hospital got selected, service functions were tailored to fit in the shell. It is just like a renovation project. As a result, the practicability of a new hospital invested with billions of yuan may be weaker than that of a renovation and expansion project.

Xi Zhu: This is quite common in China since a public hospitals' request for proposals is typically based on limited project planning and allows a very short period of time for design. Hence, the bidding turns out to be 'a beauty contest.' The approach of "building a dojo in a predetermined snail shell"[1] is how we end up with designing a new building like a renovation or expansion project.

About how design institutes design medical buildings

Xi Zhu: Of each hospital, the institutional framework, human resources, and specialized departments is unique and requires customized design from the architects to address. However, the reality is that many architects tend to use pre-developed modules for hospital design. Reasons for this phenomenon include low-level design fees, high risks of originating, and limited design time, etc. In order to catch up with the schedule and reduce workloads, designers copy pre-developed modules. As a result, hospital buildings designed by them couldn't meet practical requirements. Moreover, after making initial decisions on building scales, many architects just follow the stipulations of *Construction Standard for Hospitals* to make decisions on the seven hospital departments, without any evidence regarding the specific medical construction contents, department sizes, and sub-projects, etc. Once the projects were built and occupied, problems would begin to arise, such as the omission of sub-projects, missing of necessary rooms, and oversized or undersized rooms. These would make the costs and quality of the projects hard to control.

About the quality control in the industry

Lun Ge: As far as I'm concerned, we should sort out the processes of hospital construction management by emphasizing the process of preliminary planning and enrolling the process of POE. The existing *Construction Standard for General Hospitals* needs to be revised and updated. Evidence-based design research should be used to develop the *Construction Guidelines for Hospital Buildings*. Based on these, a system should be established nation-wide to stipulate the standards for hospital projects' quality review and evaluation, including a scientific mechanism, obligatory provisions, and basic quality assessment. Only by doing these, could we keep the quality of hospital design and construction under control.

---

[1] It's a Chinese saying that building a dojo in a snail shell. It means building something extremely complicated in an extremely small space.

Xi Zhu: It's also necessary to establish a chief architect responsibility system. During construction, a hospital client may subcontract part of the project to multiple manufacturers or teams, but the chief architect shall keep an eye on every sub-project and ensure its quality, especially the interior design. Interior designers generally pay more attention to the visual effects of design. Only with architects' supervision, could we reach a proper combination of functions and art for interior design. *Otherwise, the original architectural design concept might be modified drastically by interior designers and manufacturers, which would result in difficulties on budget and quality control.*

On the other hand, architects would develop a bad habit of doing things by haves and thus become so irresponsible that they just wait for someone else to clean up their mess in design. The chief architect responsibility system makes architects to take responsibilities for a project from the beginning to end. *Therefore, I advocate the integration of architectural design, interior design, and specialty design.*

Specialty manufacturers should focus on technical consultation and work with architects but not the decision makers on design. So, the requirements for architects are at a high level in this regard.

A hospital includes a group of systems. Each of the systems should be managed, studied, supervised, and inspected by specialty professionals. Every procedure is indispensable.

Lun Ge: We reclassified and resorted the current hospital construction procedures by stage. New stages and supplemented contents have been added to strengthen the procedures. We would like to hear your opinion on it.

Xi Zhu: First of all, I couldn't agree more on the idea of extending project planning tasks to the stage of schematic design. So that the MEP (mechanical, electrical and plumbing) engineers would be included in planning and the engineers and medical planners would collaborate on the planning and design of MEP systems. This must be completed before the information drawings were set up for design development since the MEP would in turn have a lot of requirements on building design, such as fire compartments, machine rooms, ducts and shafts, etc. These requirements should be further integrated with architectural design (Fig. 4.18).

Lun Ge: We think the preliminary planning should be done before applying for a project approval. In addition to hospital positioning, the planning should clarify every system of the hospital, including the management and operation, functional spaces, equipment, and specialty design, etc.

Xi Zhu: This is necessary. Another important point is to add a step of verification during the review of design-development documents, which help to review and decide what should be presented on the drawings. This review is critical and involves a synthetic assessment of all the professions including architecture, construction, and MEP, etc. Besides, it should contain a specialty review of the medical planning.

Fig. 4.18 The first affiliated hospital of Zhengzhou University, Henan Province

Lun Ge: We share the same thoughts again. It would be better to review the medical planning and functional layouts before the design development got approved. Is this what you mean?

Xi Zhu: Exactly. This review should also include a cost-control estimation. Once the review conclusion was made, the architect can use it to further refine and improve design documents. Furthermore, there is an issue regarding the interior decoration in hospitals. Regarding the hospital interior design, the national standards provide a list of materials but no specific construction documents on decoration. Typically, interior design is initiated upon the completion of building construction. By then, conducting an interior design would make the architecture and MEP design to be revised a lot.

To avoid the conflict, interior design should be integrated with architectural design and conducted at the stage of schematic design. The review of schematic design is critical and should include a sub-review of interior design concepts. Design drawings should respond to the preliminary planning and project positioning. Drawings should be developed in accordance with the level of investment. Investment decides the standards of design and so on. It's necessary to refine design drawings in accordance with review comments.

Lun Ge: This idea has expanded the range of what we need to do in the early stage of design. In the stage of schematic design, we should start working on interior design and medical planning. We can integrate these professional fields into one system.

Xi Zhu: All other types of public buildings follow this design principle, such as museums, hotels, schools, gymnasium, factory etc., except hospital buildings. Architectural design, interior design, and functional planning have been separated in hospital design and construction.

Lun Ge: Again, regarding project charters, is it necessary for the government or the health Bureau to organize an expert review?

Xi Zhu: A project charter should be part of the project planning and represent the outcome of the planning. The purpose of developing a project charter is more than assigning tasks and each of the professions in hospital design should have its own project charter. For instance, regarding building structure design, there should be a charter containing items such as the selected structure type and the methods to deal with vibration such as attenuation or isolation, etc. These decisions would have great impacts on building design. Another example is that there should be a charter for hospital information systems to introduce which systems should be adopted and the standards of design. These decisions would have great impacts on hospital operation and the costs of construction. So, these charters should be very comprehensive and applicable to various projects.

The weight of planning in a hospital project is more than that in a general building design. In fact, a successful project planning is even more important than a building design. The NDRC (National Development and Reform Commission) has issued certain documents indicating that 1% of the project budget should be allocated to project planning. However, it doesn't clearly indicate what to do in the phase of planning and to what extent. Frequently, hospital design and construction in China got done in a rush.

In order to speed up the processes of a hospital project, generally no one is willing to invite bids for a project planning, because few proprietors and agencies know how to do it.

Preliminary planning isn't required by the government. Construction companies and hospitals haven't realized the importance of preliminary planning for a hospital project. Only the reports on project feasibility in a fixed format are thought to be necessary. So, scientific and comprehensive preliminary planning is generally missing in hospital design.

In real estate industry, people attach great importance to preliminary planning, especially to project positioning, marketing, and return on investment (ROI). Invitation for bids in regard to preliminary planning is fairly common. Since hospital construction generally involves large number of investments, the preliminary planning should be even more important than that in real estate.

About medium-term design

Lun Ge: If a preliminary planning includes the schematic design selected via design bidding and its design development, the next design stage is then focused on construction documents.

Xi Zhu: Yes. Construction documents include multiple sections such as the architectural design, interior design, specialty projects, landscape design, etc. These design tasks should be done simultaneously with inputs and feedbacks from each other. For instance, an interior design should be integrated with the design of mechanical or electro terminals in the interior. It's necessary to establish a chief architect responsibility system to mediate the conflicts among professions. The aim is to avoid the situation that each party does things in its own way. Construction documents represent not only the design development, but also a lot of design details. It should follow the requirements set by the established construction budget.

Lun Ge: Nowadays, many interior design companies with no medical-design experience are invited to modify hospital design drawings. It has been called as 'hospital functional re-design'. Medical architects' right of design has been basically deprived by manufacturers and interior designers. Many specialty design tasks are entrusted to equipment manufacturers and thus there are blanks everywhere on architectural drawings with a note saying "To be designed by manufacturers". So, architects' work has become an empty shell and only the building facades are kept. As a result, the medical architectural system loses its integrity.

Xi Zhu: This is because architects haven't done their jobs well and interior designers have done too much. The stage of interior design was skipped in many hospital projects. On one hand, the client was unwilling to pay for interior design. On the other hand, technical and interior designs done by the architects weren't elaborated. The mess was then left to the interior designers. So, the redesign of technical functions is a make-up of the design development.

This phenomenon is due to the fact that architects aren't in charge of the design whereas interior designers are offside. The technical design of medical functions hasn't been developed well enough to provide comprehensive or accurate information for the next design stage and thus becomes a mess for interior designers to clean up. In fact, the technical design of medical functions is a supplement to architectural design in hospital projects.

Lun Ge: Only by emphasizing the chief architect responsibility system, could we manage a hospital design project as an integrated system. This is of great significance.

Xi Zhu: This is called as "architect services" in foreign countries. Architects' duty is to design and provide multiple services regarding the design simultaneously. In fact, workloads of the two parties are similar. If you sell only drawings but no services, it won't work. So, the general control of a design project is a set of consulting services that should be in parallel with the architectural design.

The professional design of a hospital is comprised of technological, architectural design, and interior design. Only senior architects who have a good understanding of all these aspects can keep everything within control.

About the stage of final inspection and project acceptance

Lun Ge: Is it necessary to add a specialty inspection on the technical design of functional flows at the stage of acceptance inspection?

Xi Zhu: The stage of final inspection and project acceptance is crucial. Currently, it involves two aspects: building inspection and operation performance assessment that is conducted a year after building occupancy.

The building inspection can be performed on site upon the completion of construction. If a building met the standards of construction, the architects then have fulfilled their duty. The assessment of operation performance would be done in a year after occupancy, focusing on the quality of construction during its warranty period. At present, hospital projects in China indeed lack acceptance inspections regarding the technical design of medical flows and could only rely on user complaints for project rectification.

Lun Ge: What we have just discussed is very meaningful. The process of building a new hospital should include three stages: the stage of project approval and design, the stage of construction documents, and the stage of construction including acceptance inspection and post occupancy evaluation. The first stage includes both documents (project proposals, project briefs, and reports on project feasibility) and drawings (schematic design and design development documents).

Xi Zhu: If these stages could be implemented, it would be a revolution in the field of hospital design and construction. The quality of hospital projects will be greatly improved.

About construction standards

Lun Ge: Next, let us discuss the existing *Construction Standards for General Hospitals*. In terms of environmental function and spatial relationship, we re-classified the seven major departments listed in the standards into ten systems. Each of the systems contains multiple functional units. In the standards, the seven departments are classified from the perspective of facility management. However, we process architectural design on the basis of functions and spatial relationships. For instance, hospital streets, commercial areas, parking lots, etc. shall be classified as a system of traffic and public service areas.

Xi Zhu: This arrangement by system is indeed specific. Associated with a definite total floor area, the area of each component in a hospital should be allocated based on its specific needs. In our daily work, we have similar thoughts but have not yet reached the level of rationality as you do. It would be awesome if hospital systems were re-categorized in this way.

A hospital system acts organically like a body system, assigning the area of each component appropriately. However, in the current standards, the distribution of department areas is not only chaotic but also unbalanced and thus misleads the design work of medical architects.

The classification of seven departments is like what happens in a beginner's design, in which each department is located in an independent building such as an outpatient building, a medical technology building, an inpatient building, an administrative building, a dormitory, etc. These buildings are connected via corridors. If an independent radiotherapy department and a large reproductive center are needed, large areas are requested in addition to the long corridors necessary for connection. However, no quotas were planned for these areas. The only way to solve the problem is to cut the size of the seven buildings to make up the areas for corridors. As a result, room areas in each department would be insufficient. Such standards are bound to mislead design.

Lun Ge: Nowadays many are still lost in such a misunderstanding. They still refer to each department as an independent building, say, outpatient building, technology building, inpatient building, etc.

Xi Zhu: Particularly, the concept of diagnostic and treatment (D&T) buildings has almost disappeared. Today we have large outpatient buildings with the functions of D&T. The integration of clinic services, technical services, and hospitalization has resulted in the emergence of outpatient complexes, D&T complexes, and inpatient complexes. To support hospital services, administration offices may scatter throughout the hospital. The true need of a hospital is patient-centered or operational-centered design concept (Fig. 4.19).

Lun Ge: Hospitals in China vary due to the weather and local cultural differences between the south and north as well as by region. It's inappropriate to make unified standards for all the hospitals. Hence, we developed both fixed and open standards. What do you think?

Fig. 4.19 The first affiliated hospital of Guangdong Medical University National Center for Respiratory Medicine

Xi Zhu: This is quite essential. For instance, hospitals in Beijing, Shanghai, and Guangzhou shouldn't be designed in the same way. These hospitals require special considerations. For instance, the standards of buildings' thermal insulation are different. In the north, it is particularly important for patient wards to face south. However, in the south, the wards may face south, north, or east, as far as not face west. Locker rooms are necessary in the northern areas, because it's warm in the interior due to general indoor heating and outpatients need spaces to storage their heavy outwears. Whereas, this is totally unnecessary in the south.

Lun Ge: The phenomenon of "dark rooms (windowless rooms)" in hospitals is quite a problem. Is it necessary to set a specific limit on the number of dark rooms? For instance, at least 70% of the major rooms should have natural lighting and ventilation.

Xi Zhu: Some consulting rooms, waiting rooms, and wards is required to have natural lighting and ventilation. It's allowed to soften the standards regarding storage rooms, mechanical rooms, and equipment rooms, etc. At present, there are no specific restrictions on the proportion of 'dark rooms' yet. It's just stipulated that over half of the wards should have natural lighting. It's necessary to formulate specific mandatory norms for this issue.

About the Post Occupancy Evaluation System—SHAPE

Lun Ge: SHAPE is a post occupancy evaluation (POE) system we have proposed, including five sections. A conclusion shall be reached regarding the quality of a hospital project by conducting a POE, such as qualified, good, or excellent.

Xi Zhu: The framework of SHAPE is proper and concise. SHAPE is well refined. There should be specific standards for rating the work of each profession who participates in a hospital project. It is similar to the star rating system for hotels. It's necessary to learn from real estate professionals on planning and learn from hotel managers on the rating system of construction. *In my view, all public hospitals should meet the standards at a qualified level and the government should formulate certain rating systems and standards correspondingly.* Design drawings shall be consistent with project budget. Meanwhile, we should specify relevant standards of each system.

Lun Ge: At present, hospitals are typically built with beautiful interior decoration. When problems show up after building occupancy, people would say that architecture is an art of regret. However, the questions are as follows: what are the unavoidable regrets, what are the regrets caused by items done improperly, and what are the mistakes that should never be made. There are no agencies who work on this type of evaluation, neither the corresponding evaluation tools, leading to a lot of problems showing up after hospital occupancy. However, POE may be offensive to many project leaders and thus hard to conduct.

Xi Zhu: This shows the traditional Chinese culture: people like a happy ending. However, design work can only be promoted via criticism and self-criticism. Once when I visited the United States, they asked me regarding my profession. I am an architectural critic, mainly criticizing myself and others. They were very surprised about that. Many people in China lack a critical spirit. To them, 'face' is more important than anything else.

Topics about the industry

Lun Ge: how do you think about the hospital designs in China, including your designs?

Xi Zhu: It is hard to tell. We are now in the situations as follows: Designs and services are frequently unqualified; consultations are deficient; the scale of general contracting isn't extensive enough.

The medical construction system in our country is sick and needs to be fixed. In the field of construction, some systems should be more flexible in order to adapt to the development of science. You have to know what's going on in the systems first, and then try to solve the problems. Now you work on the "Chinese dream" of medical system reform. We realize it is of great significance. Our current situations are following: The design is not in place; the service is not in place; the consulting effort is not enough; and the general contracting of hospital design is not deep and inclusive enough. As a result, the system would be out of control.

Lun Ge: The first purpose of my interviews is to listen to the opinions of experts regarding the results of our elaborate research done in these years. It is also a good opportunity to gather the insights of various experts, which could be formulated into a strong voice to influence government decisions and refine the systems. My job is to collect expert's ideas and link them together like a neckless of pearls. I hope the interviews will have some impact.

Xi Zhu: You've done a great and meaningful work.

Lun Ge: Nowadays, many oversea design companies have entered the Chinese market. How do you think of this phenomenon?

Xi Zhu: Oversea design companies come to China for business and profits. We should keep a clear mind in purchasing their services instead of accepting everything blindly. The hospital systems in foreign countries are different from ours, so it's impossible to copy their modes of medical services or simply buy their designs. We must not borrow it indiscriminately, because the proposals from them aren't necessarily pertinent to our domestic situations.

We shouldn't copy their concept directly without analyzing its fitness to our domestic situations. So, what are their advantages? I think they have some advanced planning concepts, design philosophies, styled designs of building appearance, and good methods of humanistic design and project management. In my opinion, if we were to do business with them, we should buy their methods of providing consulting services as well as the interior design and artistic design ideas in relation to medical materials and technologies. *I hope that oversea architects can bring in some advanced materials and systematic application technologies, etc.* It is important that we should "learn from the goodness of oversea experiences and adapt it to meet our needs", and thus "localize" it.

Fig. 4.20 Xi Zhu and Lun Ge

Lun Ge: Thank you! It was quite a pleasure that we had such an in-depth exchange of ideas in the pleasant weather and environments. I really enjoy it and it is really meaningful (Fig. 4.20).

4.6 Culture and Humanity in Designing the Drum Tower Hospital—An Exclusive Interview with Vincent Zhang

Time: January 2015—reiterated in May 2023
Location: The 815 lane, North Shanxi Road, Shanghai

Vincent Zhang: Swiss-Chinese, Registered architect of Swiss Federal Institute of Architects and Engineers (SIA), Founder of Lemanarc SA (an architecture and urban planning design office) in Switzerland, Chief Architect of the Drum Tower Hospital in Nanjing. One of the first Chinese in Switzerland's history to become a registered architect of the Swiss Federal Institute of Architects and Engineers (SIA), pioneer of the ideology of Gardenized Architecture.

Representative work: Nanjing Drum Tower Hospital, Jiangsu Province (Fig. 4.21).

Key sentences:

1. *As a Chinese architect, one should design in the context of local Chinese cultures and then consider culture integration.*

Fig. 4.21 Vincent Zhang

2. *In the entire system of Drum Tower Hospital, functions always come before forms. It reflects the ideology of rational design. The modular composition of "Medical Matrix" in design responds to service functions.*
3. *It's necessary to think about not only how to construct the entire building at a reasonable cost but also how to adapt this with the current construction technologies in China.*

Lun Ge: Design of the Drum Tower Hospital has changed the traditional style of hospitals in China and brought us a lot of surprises. As the chief architect of this hospital, you must have a rich experience in this field. What's your opinion on the ideology of hospital design? Could you talk about it?

Vincent Zhang: Based on the design of Drum Tower Hospital, I do have some thoughts and feelings regarding hospital design which I would like to share (Fig. 2.23).

First of all, a core idea I constantly practice when designing hospitals is to integrate architecture with its local history and humanistic cultures. Each hospital reflects its regional characteristic through its medical culture, making them blend to its local context and at the same time unique between each other.

The Drum Tower Hospital has a long history. It was one of the first missionary hospitals built in China since 1892 and used to be involved with the exchanges and integration of different cultures. The growth of western medicine in China was the history how foreign cultures survived and grew up on the land of China. This was quite difficult back in history. Meanwhile, no matter how western cultures have grown in China, Chinese do have their understanding of western cultures. Moreover, Nanjing had been the capital city of six dynasties in China. The Drum Tower had witnessed numerous historical events. The Drum Tower Hospital belongs to its place as well as to its local culture.

In my opinion, there is no hospital built in a single style. Just like other buildings which have the same "international style" would never be identical. Whereas, in the phenomenon of 'globalization', a lot of hospitals look quite the same, as machine-made. It is somewhat scary. *As a Chinese architect, one should design in the context of local Chinese cultures and then consider culture integration.* From this historical and cultural perspective, what differentiates the Drum Tower Hospital from other hospitals? What are the unique Chinese cultural characteristics to build a hospital? How to integrate the rationality of western medicine and church cultures in the design? These types of questions were studied during the design process.

Secondly, I'd like to talk about the combination of design forms and building functions. *In the entire system of Drum Tower Hospital, functions always come before forms. It reflects the ideology of rational design. The modular composition of "Medical Matrix" in design responds to service functions.* In other words, medical functions are the basis of how these departments are arranged, how they interact with each other, and how these congruent departments develop in the future. After responding to these fundamental questions, the design forms gets developed. I was lucky to be able to dialectically integrate the forms and functions well rather than picking a "symbolic" building image for this hospital and then fill in the functions inside the building, as many other hospitals do (Fig. 4.22).

Lun Ge: You are an architect who has enriched the implication and experience of hospitals. Also, you've talked about the rational design of buildings and your thoughts regarding the relationship between forms and functions. Could you share with us more details in this regard?

Vincent Zhang: The combination of design forms and functions can be quite interesting and artful. People might think that the Drum Tower Hospital has an attractive and ingenious facade in terms of design. Actually, it's also effective in saving energy and preventing possible suicides and so on. Regarding the local climate, Nanjing is

Fig. 4.22 Courtyard of Nanjing Drum Tower Hospital, Jiangsu Province

one of the four hottest cities in China. Therefore, it's of great importance to solve the issue of exterior shading. Moreover, we need to consider how to reduce sound pollution in the downtown area as well as how to ensure natural daylighting in the interior. Finally, we chose grounded glass for the windows. This material diffuses daylight, leading to soft daylighting and well-lit interior, reducing the amount of energy needed for artificial interior light. *These design factors were intended to meet patients' genuine physical and psychological needs during medical care, instead of being eye catching which wasn't our intention.*

In fact, we still have to consider a lot of practical issues such as building budget. *It's necessary to think about not only how to construct the entire building at a reasonable cost but also how to adapt this with the current construction technologies in China.* We used the method of unitary approach. At the same time, with this method we could ensure the quality of certain components when they were under fabrication.

In addition to quality control, combining design forms with functions is very important. The reason that we chose ground glass was not only for energy saving but also for stain-resistance. Many people think that ground glass look dirtier than normal glass, which is not true.

Regarding this unawareness, we decided to collected different types of evidence and use this to convince the hospital to approve our design. We chose a type of one-side ground glass, of which the external side is frosted but the internal side is not. In order to show how ground glass resisted stains, we compared the stain-resistant ability of clear and ground glass. Both types of glass were clearly scarped for comparison. After splashing dust and ashes on them, it was quite obvious that the pane of clear glass looked dirtier straight away. Moreover, smear strips remained when it got wet in rainy days, but this was not the case for the pane of ground glass. This told us that people may misunderstand things habitually. Only by integrating design forms with functions, we could comfortably convince our clients. If we look back, the decisions on the facade design and materials were pretty good. *Focusing on natural ventilation and daylighting, this design allows users to raise plants on the windowsills, which addresses both the physical and psychological needs of patients. On the other hand, it has proved its efficiency on energy saving and contribution to cost control as well.*

Continuous communication with the client is crucial during the design. I think President Rong Ma at Drum Tower Hospital is quite extraordinary, and I very much like people like him. Although they major in western medicine and their opinions on design could be quite different compared to an architect, nevertheless they value logicality. This is extremely similar to the rigorous of architects with their design. As a result, when we discussed or analyzed design issues with hospital representatives, our ways of thinking were very much the same as theirs. Thus, a basic platform for communication was established on which we could discuss and work out solutions together, by eventually making agreements about the design to satisfy both parties (Fig. 4.23).

Lun Ge: You mentioned design in the context of cultures. Could you talk about the relationship between architecture and cultures as well as design originality?

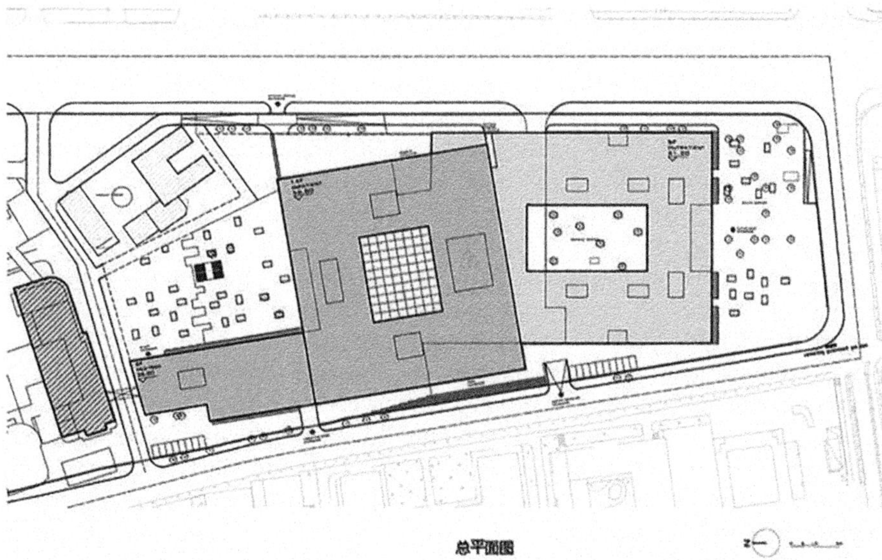

Fig. 4.23 General plan of Nanjing Drum Tower Hospital, Jiangsu Province

Vincent Zhang: *Architecture is a carrier of culture.* It's pointless to talk about medical culture before an architect completes the design of a building. It's weak to discuss about the medical culture of a hospital in generic terms.

Nevertheless, it's an architect's duty to translate these 'adjectives' into reality. Hence, architecture is a carrier of culture. In these kinds of buildings, every healthcare professional has a sense of identity, dignity, and honor about their role. This is of great significance. In the meantime, architects achieve their personal value.

In terms of spatial planning, the enclosed-shape of layout is also designed to address the actual needs of this hospital. On one hand, regarding medical spaces with the same capacity, the preference should be given to low-height buildings as vertical transportation has a strong influence on travel flows. For instance, there would be queues of people waiting for elevators. On the other hand, we should avoid designing public spaces that can't be controlled. *Based on traditional thoughts on building design, many consider the ground floor as a grand square. However, we consider that the ground-level open square would become a market place filled with people and cars.* The public space tends to be a mess and out of control in this regard. Instead of choosing an out-of-control space, we prefer to build a calm and therapeutic space. This reminded me that traditional therapeutic spaces in medieval times was actually a kind of religious space, which were mainly enclosed courtyard spaces. Only in this way could people calm down in the downtown area. Therefore, the fundamental values and methods of developing therapeutic architecture have remained the same, making the starting point of design based on the needs of the functions.

Lun Ge: During our visit to the Drum Tower Hospital, we also felt the originality and ingenuity of you as an architect. Could you share more about your design philosophy and the details?

Vincent Zhang: *Personally speaking, in fact, I really hope that hospitals could give people a tranquil sense of religiosity.* When getting sick, people often grow relatively weaker than usual and hope to receive help from others. Take the small platforms around the atrium in our design for example: functionally, they were designed for professors and students to hold informal meetings, similar to the stages in medieval churches for choirs. Doctors can feel free to use these platforms at their convenience. *I think the dimension of design is humanistic.*

Besides, I also pursue meeting the needs of hospital environments. It's said that the Drum Tower Hospital looks like a museum and an art gallery and actually, I take it as pure coincidence. There are no specifications on how an art gallery should look like, nor any requirement on how a hospital should look like.

At the time, the managers in charge specified that hospitals should be built like a "hospital". What does 'looking' like a hospital mean? If it means we should build the new hospital as those built decades before, we may repeat building environments with a sense of dullness and stress, which is quite scary. People also say "being like a hospital" meaning to be ultra-luxury such as using expensive materials like marble. If this is the case, it involves a diverse set of other associations.. What we intended to design was a hospital and not a club. Looking at a hospital building or the materials may not demonstrate design concepts. "Being in a hospital" is, in fact, just a strained interpretation made by people with limited understandings about of hospitals. *Therefore, the core of hospital design should be addressing functional needs and integrate design forms with functions.*

For further details, let's give two more simple examples. One is about the design of the ceiling, which is a very special design process for me. At that time when we were dealing with this kind of ceiling, there were no example to do gauze net ceiling in domestic buildings. Meanwhile I didn't want to simply copy the western suspended ceiling with a rolling seine. This is like the fence on the side of the highway, which owns both light weight and thickness. In the design process, I originally specified that the net height could not be less than 2.7 m, but after the engineers took the structure and ventilation into account, only 2.6 m was left. Meaning that 10 cm of the space was lost. In the original design, the lamp should be rubbing the net sideways in the scope of 10 cm, by hiding the light source, you can only see the light inside the net, aiming to see the whole ceiling shine evenly. This can only be achieved with the scope of 10 cm. For this, the only solution was to increase the thickness of the net, we went to the factory and continually experimented until we found 1 mm aluminium foil. The next step we had to fix was the packing case with aluminium foil to fit the net with its required thickness. After this specification was solved, the inside light softened again.

Another example is the façade design. Actually, we took three years to design the whole hospital. In these three years, we've finished all the details of the interior execution drawings, including all the details for the façade design. Normally domestic architects don't do the façade design and will ask a façade company or façade consultancy company to take this job, to then ask the façade construction company to do it. The façade construction company will then refer to their previous work to then make it directly. At the very beginning, we also tried to do it this way, but at the end the key node of the façade design showed nothing and we were unable to proceed further. So, we designed all the detail from the beginning, took these detail drawings to discuss with the façade company and then the construction company. And We did it 6 times just for the big sample of the facade.

The external facade of the hospital has three layers. It includes a concave window system, oriel windows, and an outboard layer. Concave windows are applied with clear glass, whereas ground glass is used for oriel windows. The outboard layer is built with aluminium orifice plates. *If the transmittance indexes and refractive indexes of the three layers vary, it would be impossible to achieve the intended design of layered transparent paper floating one another and the soft illusion of traditional Chinese window paper.* However, the calculation methodology for measuring refractive indexes, reflective indexes, and the transmittance of aluminium plates differed from industrial standard. Hence why we had to measure these numbers on factory sites and then conduct pilot studies. Eventually, we managed to develop the existing appearance of the hospital. Actually, the suspended ceilings and curtain walls were also built by integrating design forms with hospital functions in the context of practice. Laborious as it was, it was well worth it as well.

Lun Ge: I admire the spirit of Mr. Zhang and his outstanding performance of the Drum Tower Hospital. So, could you tell us about your opinion in terms of SHAPE-post-evaluation system?

Vincent Zhang: In terms of traffic safety, vehicles enter the Drum Tower Hospital through the underground level. There were only a few cases of this kind of design in China at that time. Many hospitals were learning from the mode of Singapore hospitals. Besides, most designers still insisted that hospital buildings should be high; the ground level should be spacious; and there should be a huge square. *Comparing with these designs, our design of the Drum Tower Hospital has shown higher levels of safety. There are no vehicles on the ground level and thus no conflicts between vehicles and pedestrians.* In this regard, I think that traffic safety has been improved making pedestrians walk freely without any vehicle interference.

Regarding safety, some people back then criticized our design for building no walls. Yet, no one mentioned about it since then. The building itself serves as a good fence for the hospital. The public space inside the building is a series of "courtyards" surrounded by walls. Besides, as a public product for the city, the building itself is supposed to integrate with the city rather than being a fenced traditional "compound".

Upon humanity, the design of the Drum Tower Hospital is prepared to become international in the up coming future in terms of environmental lighting, ventilation, and the reservation for doctor offices. *In addition, we designed patios and courtyards, which responded to the needs of humanistic environments.* In the design, we included a registration office and a pharmacy on each floor. Moreover, we built resting areas in the inpatient department for patients and their family members. These functional factors were quite progressive at the time of the project proposal. However, they appear to be quite necessary now.

When designing the Drum Tower Hospital, I intended to build it as an "island". It is located in the downtown area and should be different from surrounding buildings in terms of building style, material, and color. Taking a closer look at it, you could see the gap between the building and the surrounding sites. The place by which people entered the hospital was constructed like a bridge. In fact, I wanted to make it an "isolated island" in the city, which is actually a space connecting "this world" (the real world) and "that world" (the underworld). We were "born" here. We were to "die" here. We also stayed here when we were helpless.

Actually, in ancient European, the civilization of medical treatment was titled "islands." Hospitals transported patients to an "island". Though this originally was made to segregate the sick ones, people then gradually felt that therapy was highly valued and connected the two sides of the river named life. Therefore, "humanity" never meant to be simple, vulgar, or utilitarian. Instead, it is the belief and one of the basic needs of people. We intended to borrow this type of sacred approach to promote the awareness of both patients and doctors in valuing "medical" activities. However, in reality, it's hard for doctors in China to have such awareness. Due to the large number of patients in China, they may need to treat a patient every three minutes, which may feel like a production factory. In fact, they may have forgotten their sense of purpose during their medical services. We hope to bring it back through our future designs.

Regarding the functional spaces, what's interesting in our design is the integration of the inpatient and outpatient departments. Usually, the two are separated, but their integration can actually benefit us in many aspects. This design could facilitate services in a relatively more centralized manner, forming a relatively clear sign. Furthermore, to a great extent, it improves the integration between the central pharmacy and every inpatient floor. The work flows are shortened and a lot of money is saved. Besides, the annular series design provides independent systems for both the doctors and patients, leading to a better operation efficiency. *Future hospitals are supposed to be a neural network. In fact, we have implanted this network in our design for staff members, which is free from the influence of patient flows.* Patient flows are served in the core area of the hospital, whereas the spaces for the staff are located in the surrounding areas of each core area. Such design is prepared for future development and the practicability behind it.

In terms of physical environment, it's true that there have been some adjustments made on lighting design. For instance, originally, lighting for all nurse stations were required to be warm, but cold lighting was added instead. This may have weakened the desired warm feeling and atmosphere. It is a shame.

Regarding efficiency, I talked with the staff members. They like the spatial layout of the new building. It provides the doctors an independent and quite undisturbed working area. With this, they worked better in their duties, creating an overall improvement in efficiency and satisfaction.

Fig. 4.24 Vincent Zhang and Lun Ge

Lun Ge: Thanks for your wonderful analysis. It's really not easy for an architect to achieve such pursuit. Thank you again (Fig. 4.24).

The manufacturer's authorised representative in the EU is Springer Nature Customer Service Centre GmbH, Europaplatz 3, 69115 Heidelberg, Germany. If you have any concerns regarding our products, please contact ProductSafety@springernature.com

Printed and bound by CPI Group (UK) Ltd, Croydon, CR0 4YY
30/09/2026
02225265-0018